# Governing Transformative Technological Innovation

To RHD (Bob) Phillips, Dad,
who showed me the importance of asking questions

# Governing Transformative Technological Innovation

## Who's in Charge?

Peter W. B. Phillips
*University of Saskatchewan, Canada*

**Edward Elgar**
Cheltenham, UK • Northampton, MA, USA

Published by
Edward Elgar Publishing Limited
Glensanda House
Montpellier Parade
Cheltenham
Glos GL50 1UA
UK

Edward Elgar Publishing, Inc.
William Pratt House
9 Dewey Court
Northampton
Massachusetts 01060
USA

A catalogue record for this book
is available from the British Library

**Library of Congress Cataloging in Publication Data**

Phillips, Peter W. B.
Governing transformative technological innovation : who's in charge? /
Peter W. B. Phillips.
p. cm.
Includes bibliographical references and index.
1. Technological innovations—Management. 2. Technological
innovations—Economic aspects. I. Title.
HD45.P519 2007
338′.064—dc22 2006034058

ISBN 978 1 84720 237 6

Typeset by Cambrian Typesetters, Camberley, Surrey
Printed and bound in Great Britain by MPG Books Ltd, Bodmin, Cornwall

# Contents

# Abbreviations and acronyms

| | |
|---|---|
| AAFC | Agriculture and Agri Food Canada |
| ACRE | Advisory Committee on Releases into the Environment (UK) |
| APHIS | Animal and Plant Health Inspection Service (US) |
| ARPA | Advanced Research Projects Agency (US) |
| ART | accountable, responsible and transparent |
| BIO | Biotechnology Industry Association (BIO) |
| *Bt* | *Bacillus thuringiensis* |
| CAC | Consumers' Association of Canada |
| CBAC | Canadian Biotechnology Advisory Committee |
| CBD | Convention on Biological Diversity |
| CEC | Commission for Environmental Cooperation |
| CFIA | Canadian Food Inspection Agency |
| CGIAR | Consultative Group on International Agricultural Research |
| Codex | *Codex Alimentarius* |
| CRADA | collaborative research and development agreements |
| CSIRO | Commonwealth Science and Industrial Research Organization |
| CSO | civil society organization |
| CSR | corporate social responsibility |
| EPA | Environmental Protection Agency (US) |
| EU | European Union |
| FAO | Food and Agriculture Organization of the United Nations |
| FDA | Food and Drug Administration (US) |
| G7 | Group of Seven leading industrial democracies |
| G8 | G7 plus Russia |
| GATT | General Agreement on Tariffs and Trade |
| GDP | gross domestic product |
| GLC | gas liquid chromatography |
| GLP | good laboratory practices |
| GMP | good manufacturing practices |
| GMOs | genetically modified organisms |
| GPT | general purpose technology |
| GRP | good research practices |
| HACCP | hazard analysis and critical control programs |
| HEAR | high erucic acid rapeseed |
| HPV | human papillomavirus virus |

| | |
|---|---|
| IANA | Internet Assigned Numbers Authority (US) |
| ICANN | Internet Corporation for Assigned Names and Numbers (US) |
| ICT | information and communications technology |
| IEC | International Electrotechnical Commission |
| IPPC | International Plant Protection Convention |
| IPPM | identity preserved production and marketing |
| IPR | intellectual property right |
| ISI | Institute for Scientific Information |
| KBV | knowledge-based view of the firm |
| MRC | Medical Research Council (UK) |
| NAFTA | North American Free Trade Agreement |
| NBF | new biotechnology firm |
| NCE | National Centers of Excellence (Canada) |
| NGO | nongovernmental organization |
| NIH | National Institutes of Health (US) |
| NRC | National Research Council (Canada) |
| OECD | Organisation for Economic Co-operation and Development |
| OIE | International Office of Epizootics |
| PBV | product-based view of the firm |
| RBV | resource-based view of the firm |
| REB | research ethics board |
| SNA | social network analysis |
| SOP | standard operating procedures |
| SPS | Sanitary and Phytosanitary Agreement of the WTO |
| TBT | Technical Barriers to Trade Agreement of the WTO |
| TRIPS | Trade-Related Aspects of Intellectual Property Rights (WTO) |
| UK | United Kingdom |
| UN | United Nations |
| UNDP | UN Development Program |
| US | United States of America |
| US PTO | Patent and Trademark Organization (US) |
| VCC | venture capital corporation |
| WHO | World Health Organization |
| WTO | World Trade Organization |

# Preface

My alma mater, the London School of Economics and Political Science, displays on its crest the dictum *rerum cognoscere causas*, which literally translated means 'to know the causes of things'. I have trained at points in my career and worked professionally as an economist, a political scientist, a policy analyst, a teacher, a scholar and a consultant and have become increasingly concerned at the narrow research approach many of my colleagues and peers have adopted in their study of technological change. They generally begin from a disciplinary perspective, which limits their ability to follow the problems they are examining and impedes their capacity to identify true, causal relationships. I have deliberately adopted an interdisciplinary approach in this volume in an effort to find where there are strong, causal explanations for events and where there are competing or contested explanations or, at times, where there is a paucity of explanation.

Ultimately, one question underlies much of the angst in this post-industrial era dominated by globalization: who's in control? We have begun to realize that the state is not the sole or even dominant governor in many fields. Similarly, although the market is in the ascendancy, it quite explicitly does not aspire to govern large parts of our system. Recently, we have begun to look to a new set of civil authorities, sometimes dubbed the third sector or not-for-profit sector of civil society. Although some groups have power and authority, they frequently do not mesh well with the delimited competence and capacity of the other two, more traditional governors. This book is seeking to understand, in the context of transformative innovation, where these three governors work well separately, or together, and where they don't.

Aristotle in *De Poetica* asserted that 'the greatest thing by far is to be a master of metaphor . . . since a good metaphor implies an intuitive perception of the similarity in the dissimilar'. Each of the chapters of this book starts with a metaphor of the problem to be investigated. Many of the practical examples and problems are drawn from my experiences related to transformations in, around and beyond the food and natural environment. Although many commentators and analysts think that the food and farm system is antiquated and unimportant, agri-food activities continue to represent the single largest identifiable economic effort in the world. About half the world's population, roughly three billion people, rely on agriculture for their main income, with about one billion mainly subsistence farmers. Similarly 40 per cent of the world's population use renewable fuels (wood, charcoal, straw and biomass) for at least some of their energy needs and about one billion depend directly on fishing for a large portion of their protein.[1] Even in advanced industrial economies, the complex supply chains spanning from Petri dish to dining table are one of the most significant economic contributors to the economy and society.

This book is the product of a dedicated sabbatical leave that allowed me the luxury of time to read widely, to meet key individuals to discuss my thoughts and to ponder, synthesize and focus my ideas from 25 years of scholarship and professional practice.

Many people helped me test my ideas. My graduate students – including Jeremy Karwandy, Lisa Jategaonkar, Tara Procyshyn, Camille Ryan and Stuart Smyth – have both acted as sounding boards and provided me with insights into the literature and the issues. I would like to thank all of my colleagues, both at the University of Saskatchewan and around the world for listening to my thoughts and challenging me to justify my ideas more clearly and succinctly. Finally, I must acknowledge my friends and family who have put up with my obsession to undertake this work.

I would like to thank the University of Saskatchewan, my colleagues in my various departments and colleges and my students for allowing the opportunity to leave for a year. I gratefully acknowledge my friends and colleagues who offered me sanctuary and collegiality during my sabbatical: the Department of Sociology and BIOS at the London School of Economics and Political Science; the Agricultural Directorate at the OECD; the Department of Political and Social Sciences at the European University Institute, Florence; and the Institute for Advanced Studies and the School of Agricultural and Resource Economics at the University of Western Australia, Perth.

**Peter W. B. Phillips**
**Saskatoon, Canada**

## NOTE

1. Homer-Dixon (2001).

# PART I

# The challenge

# 1. Introduction

The next time you consume a french fry, potato crisp, boiled, baked, mashed or scalloped potato, gnocchi, vichyssoise, perogie or pre-cooked fish, beef or poultry, stop and think about how that food got to your plate. Each of those foods has a component derived from potatoes. On an average day, seven out of every eight people in developed countries such as Australia, Canada, European nations and the US consume one or more food products containing potatoes or potato ingredients and one in six consumers in developing countries eats some potato daily. Many consumers will also drink alcohol, such as vodka, produced from potatoes or eat meat from animals fed on potato pulp feed.[1]

In addition to food uses, potatoes are also a major source of industrial starches. The starch extracted from potatoes through industrial wet milling, refining and drying is valuable to a variety of industries, including the paper and board industries, especially for paper/pulp surface sizing. Potato starch and its derivatives are also used for textiles, adhesives, surfactants, polymers, pharmaceuticals and cosmetics. There have also been efforts to use potato fibrous materials as bulk filters for plastics.

The humble potato would hardly seem to be an auspicious example of how to examine global transformations caused by converging technologies. To many it will seem to be a timeless and never-changing food, reserved for the table of the poor and oppressed. Economists somewhat derisively use the potato famine in Ireland in the 1800s to illustrate the Giffen paradox, where demand for potatoes rose even though potato prices jumped following the potato blight (which is just the opposite of what microeconomic theory would predict). The potato market exhibited this pattern because potatoes made up such a large portion of the diet that the higher prices radically lowered the purchasing power of most peasant farmers, causing them to reduce demand for other higher-priced sources of protein and instead to eat more potatoes. For the most part, the potato is viewed with disdain both within society and literature.

Nevertheless, the potato offers interesting insights into the effects of converging technologies. Over the past decade it has been the subject of increasing enthusiasm, investment, tinkering, change and debate. In the first instance, the genetic base for the potatoes we eat has been changing rapidly, as molecular-based breeding techniques have enabled seed developers to create new cultivars that meet the increasingly diverse needs of the table, the chip fryer, the processor and industry generally. This has yielded an increasing array of traditionally bred cultivars. For example, more than 300 varieties were registered for cultivation in Canada in the 1990–2005 period, an average of about 20 new varieties each year.

Much of the new research and development has focused on changing the production system for potatoes. Commercial production of potatoes is often fraught with risk. Two key problems are the Colorado potato beetle, which eats the foliage and stems of potato plants, and two types of virus that, when transmitted by aphids, cause leaves to curl and the stems to become woody. Extensive infestations of either pest can reduce yield, as a smaller leaf surface area limits the plant's ability to produce and store nutrients, which reduces the number and size of the tubers. It is possible to control both problems with insecticides. An estimated 34 per cent of total insecticide in the potato sector is used to control the Colorado potato beetle, more than is used for any other pest. One difficulty for farmers is that the Colorado potato beetle has developed resistance to a number of the pesticides used, including an array of highly toxic and persistent arsenicals, organochlorines, carbamates, organophosphates and pyrethroids. Although aphids can be controlled by chemicals, the window of opportunity between an infestation and the concomitant spread of the virus can be as little as 24 to 48 hours, which makes it tricky to control.

In response to this problem, Monsanto Company, headquartered in St Louis, MO, sought to use new biotechnologies, including transgenic gene splicing, to develop cultivars which resist the beetle and viruses without reliance on chemicals. The company developed and introduced into Canada and the US four transgenic types of potato: two Colorado potato beetle resistant varieties (using a gene called Cry3A extracted and purified from a naturally occurring soil bacterium called *Bacillus thuringiensis*, *Bt*); a Colorado potato beetle and potato virus Y resistant variety (using a gene extracted from the virus); and a Colorado potato beetle and potato leafroll virus resistant variety (using a gene extracted from the virus).[2] The *Bt* acts as an insecticide, selectively binding to parts of the beetle's gut, causing paralysis and ultimately leading to death, and the two virus genes act to provide immunity to the plants so that the farmers need not worry about defending against the aphids.

A number of other research efforts are seeking ways to change the composition and function of potatoes. The processed food sector, in particular, wants potatoes that have lower carbohydrates (one variety currently available has 30 per cent lower carbs than competing varieties) and potatoes that are denser and have less water, so that when fried absorb less fat from the frying oils. Meanwhile, others have been examining whether potatoes might provide a higher-value service, acting either as a storage or production system for pharmaceutical proteins.[3] One Canadian group has transgenically modified a potato to produce serpin proteins, which would then be purified for commercial or pharmaceutical use.[4] Others are working on making potatoes part of the vaccine system. As early as 1986, research showed that mice showed oral immunogenicity to the Norwalk virus after consumption of transgenically modified potatoes.[5] By 2000, some human trials had shown immune responses. Potatoes are also being used to express vaccines for hepatitis B and E, and the human papillomavirus virus (HPV is linked to cervical cancer).[6] The Boyce Thompson Institute for Plant Research of Ithaca, New York, for example, in the late 1990s undertook to transgenically modify a potato to insert the vaccine for Hepatitis B, so that people might simply be able to eat a potato or a processed potato product to get the vaccine dose. Others have looked at using potatoes as a cheap, efficient, safe storage vector for pharmaceutical proteins in developing countries, where the risk of failure in the cold storage of vaccines is higher.

## WHO'S IN CONTROL?

Many of us, when confronted by the list of things others want to do to the potato, are likely to ask 'who is in control?' Who develops, produces, processes and handles these products and who is responsible for owning, controlling, managing and assessing their contents, weight, safety, nutrition and economic, social and environmental impacts?

The simple approach is either to look to the last one who visibly handled the product – for many of us that would be a retail food store such as Carrefour, a processor such as McCain or a fast-food outlet like McDonald's – or to the nearest government with an agency that appears to be responsible for agriculture or food safety.

The reality is that finding out who is in charge or responsible for any given product anywhere in the world is a challenge. Looking for who is in charge, or to put it another way who governs the product, is difficult because there is a complex and dense layer of actors and agencies that govern the final structure and function of almost any product available in the world today. Though potatoes represent a relatively simple example, they can be used to illustrate the point. Economists such as Herbert Simon or Paul Seabright suggest the solution to finding out 'who's in charge' might be to follow the transactions. Simon has posited that if transactions emitted light or heat, a special sensor from space would show a widely different array of arrangements than we see through our traditional geopolitical lens.[7] Seabright went further in his recent book entitled *Company of Strangers*, describing who was in charge of coordinating the availability of a blue cotton shirt he purchased. He used the shirt to illustrate the marvel of impersonal, uncoordinated optimizing individuals operating within markets and how they delivered a shirt on time and in the right place to satisfy his wants. Both, however, miss the point. Each of their investigations looks at transactions, and not at the underlying norms, conventions, arrangements, laws, communities and systems that enable those transactions to occur. In the absence of some institutional arrangements, individuals cannot or will not efficiently trade, and the marvel of the market will fail.

This book investigates the names, faces, institutions and actors behind the markets and looks for an analytical and causal explanation for why at times it can be a public entity that takes responsibility, at other times the market suffices and, increasingly, purpose-built (or quite often serendipitously available) self-organized groups will emerge.

With established markets, such as for the well-known french fry, there seldom are any major concerns or questions about who is in charge. We either don't think about who's in charge, or, if we do, we marvel, like Seabright, that in spite of an absence of any apparent overseer, we generally get exactly what we want, when we want it.

The difficulty in sorting out control, and the costs of not having any single actor obviously tasked with responsibility, inevitably rises when new technologies are proposed, especially markedly different technologies. Despite the verbal deluge of buzz words used to describe the advent of new technology and the attendant consequences of change – radical, disruptive, transformative, major, revolutionary – the underlying premise is that a new product, process, market or organization creates new opportunities that offer both benefits and costs. Truly new ideas are most likely also to pose the greatest uncertainty – we not only don't know how bad (or good) something might be, but also we are not even sure what the problems or opportunities arising from it might be.

In the face of uncertainty, we look to known authorities to take control and responsibility. Finding out who is likely to take control and what they are likely to do is challenging. To illustrate the point, it is worthwhile exploring the pressures on the lowly potato.

The conventional potato, without any transgenic enhancement, is governed by an array of actors that spans the globe. Finding who is really in control is a bit like peeling back an onion. Individual actors seldom control much. Most just have a specific function or aspect of the product that they control and then rely upon others to govern other aspects, either through formal arrangements or through tacit understandings (often not even mutually acknowledged).

Let's start at the table and peel back the layers of control. The potatoes you purchased can have come into your house whole (and you will have prepared them to your taste, using a variety of ingredients and methods). Many of those potatoes, even if they aren't bagged, are branded varieties that are aggressively marketed to differentiate them from common table varieties. These branding programs are grounded on domestic law and are either the property of individual producers or collective groups (such as the Idaho Potato Commission). If so, in most parts of the world you would have purchased them from a vendor, either a store or a street marketer. Alternatively, you might have purchased a frozen, ready to reheat packet of french fries, which are marketed as a branded product. Apart from any firm specific procedures and practices of the retailer (some which are set collectively within the industry, such as by the Retail Council of Canada, the Grocery Manufacturers' of America or the British Frozen Food Federation), food handlers are regulated by local, provincial, state and national laws about food safety and marketing. Many of those rules are often coordinated internationally and audited by third-party assessors. The grocers are also linked closely to the wholesalers, who have a similar set of rules and procedures to which they conform. Given that many potato products are shipped long distances and frequently across regional or national borders (for example, McCain has a large share of the German french fry market but does not have a processing operation in Germany), a wide range of rules may apply. National import and export rules and a wide array of commodity-specific and general trade agreements (such as the North American Free Trade Agreement, the European Union Common Agricultural Policy and the World Trade Organization Agreement) can have effect. A series of international product, sectoral and institutional standards – such as supply chain hazard analysis and critical control programs (HACCP), sectoral International Standards Organization (ISO) standards and *Codex Alimentarius* standards – may come into effect. Moving one step further back, processors balance an array of formal government regulations related to health, safety, the environment, worker safety and industrial strategy with their corporate strategies (related to brand value and market share) and any attendant arrangements required either by their buyers (such as McDonald's or Frito Lay) or their suppliers (legally mandated potato marketing boards and other private marketing bodies).

This brings us to the farmer, who often sees his or her lot as being squeezed by the constraints of their suppliers and the demands of their markets. Each, in turn, is increasingly proscribing the on-farm production of potatoes, specifying the bundles of inputs that must be purchased and used together (including seeds, chemicals and finance) and often establishes precise production targets for volumes and qualities, including both biochemical attributes and production and processing methods.

Moving further back in the supply chain, we come to the input providers, who market their goods or services directly or through wholesalers and sometimes are vertically connected with downstream processors and retailers. They also face an array of laws, industry rules, corporate imperatives and sectoral undertakings.

The investigation gets especially interesting when we examine the seed sector.

As noted, there has been significant interest in altering the performance of potatoes. The companies, universities, research institutes, governments, foundations and collectives who fund and undertake the basic and applied research on potatoes have an array of formal and informal rules they are constrained by and, depending on the technology and sector, have formed new institutional arrangements to manage the research enterprise. For well established products, such as the industry standard Russet Burbank variety, the underpinning knowledge and data are codified and documented in the academic journals, extension services manuals and regulatory systems.

When a new product emerges from a research effort, then, depending on the nature of the new or changed attributes, some or all of the foregoing control systems need to evaluate and decide on whether or how to adapt and adopt the technology. This process is far from linear. Given that many new traits will not fit within the competence, jurisdiction or interest of many of those currently controlling the supply chain, new institutions may be needed. Furthermore, it is likely that some changes may interest and engage a whole new set of actors and participants, frequently ones who are disenfranchised from the current control system.

Let's consider for a moment the introduction of transgenic potatoes in the late 1990s. In the mid-1990s Monsanto Corporation accessed basic and applied science from the academic community and mobilized a team of in-house and contract researchers to adapt its proprietary *Agrobacterium tumefaciens* transformation system (using a kanamycin-resistance gene linked to a constitutive promoter) to develop a range of insect and viral resistant potato varieties. Before they could even consider entering the market, they had to undertake a three-year regulatory review, beginning with confined field trials simultaneously in Canada and the US (all of the varieties were derived from male sterile Russet Burbank potatoes, which facilitated approval from the national regulators). In Canada, the Canadian Food Inspection Agency (CFIA) and Health Canada and in the US the Food and Drug Administration (FDA), the Environmental Protection Agency (EPA) and the Department of Agriculture Animal and Plant Health Inspection Service (APHIS) evaluated the traits for human safety, safety as an animal feed and environmental impacts, using an array of established, new and negotiated procedures drawn from academia, the regulatory system, Monsanto, commercial labs and international conventions (over time the scientific process of evaluation has been formalized and codified through a variety of international agreements and conventions). As part of the approvals, both countries required Monsanto to implement integrated pest management plans (involving refugia) to ensure that the target insects did not develop resistance to the new traits.

Once the varieties were approved, Monsanto trademarked the new cultivars as NewLeaf® products and established NatureMark Potatoes to market the products. Farmers initially seemed excited by the new varieties, as they worked to lower their use of insecticides and chemicals and thereby improve their profitability. In the spring of 1999 farmers and Monsanto were enthusiastic and optimistic about the future of this new technology and were looking forward to a cascading array of new production possibilities.

The wheels fell off the product in the fall of 1999. A number of seemingly disconnected events combined to raise a firestorm of concern in key markets for potatoes and ultimately led to the withdrawal of the product. Two events in particular mobilized a diverse group of people and institutions to work against the product. First, Dr John Losey's work at Cornell University on the effects of *Bt* corn pollen on Monarch butterfly larvae, published in *Nature*, generated significant media hype.[8] The second challenge was the controversial research on the food safety properties of *Bt* potatoes conducted by Dr Arpad Pusztai at the Rowett Institute in Scotland.[9] Pusztai conducted a series of tests on rats using GM potatoes and concluded that the consumption of these potatoes impaired the immune system of the rats. After he announced his findings on national TV in Britain, the public debate about this new technology ballooned. Others have examined the details of this story; suffice it to say that by 2000, retailers, processors and fast-food outlets were increasingly concerned about the technology.[10] Regardless of the assertions by Monsanto, the industry, the food processing industry and most scientific bodies that these products were safe, one by one different parts of the industry withdrew the product. The first to refuse the technology were British retailers, who indicated they would not sell the products. This was followed shortly afterward by Frito Lay announcing it would not use GM potato varieties and then, almost simultaneously, McDonald's and McCain announced they would no longer buy or sell products with GM potatoes. Ultimately, Monsanto withdrew the potatoes from Maine and Prince Edward Island, offering technical help and support to producers to identify and remove GM seed potatoes from their operations.

In this case, the convergence of new technologies in the context of the global potato market was socially unsuccessful. The reason is buried deep in the governing structures that underlie the potato system.

## STRUCTURE OF THE BOOK

In the rest of the book I examine the deep governing structures of transformative technology and innovation in an effort to identify which actors can be expected to act when and under what conditions and to what effect. Ultimately, it is not a question of finding a single entity that has control. Rather it is a matter of finding how a complex system of institutions and actors operates and how they can be made to enhance economic and social welfare. The book is not really about potatoes, but instead will examine an array of examples where converging technologies have created transformations and governing challenges. All products, whether as simple as timber produced for a building or an egg for our omelet, or as sophisticated as a replacement transmission for our automobiles or the Apple iPod, involve complex control and governing systems.

Part II of the book grounds the debate and analysis, providing assessments of the concepts of transformative technology (Chapter 2), innovation (Chapter 3) and related governing challenges (Chapter 4).

In Part III I examine the various tools commonly used to examine governing systems, including overarching paradigms for governing transformative innovation (Chapter 5) and the models and taxonomies used to investigate governing via the state (Chapter 6), the market (Chapter 7) and civil authorities (Chapter 8).

In Part IV I apply the paradigms, tools and methodologies to current transformative changes to investigate the challenges of governing in practice: first, the systems and authorities that define knowledge (Chapter 9); second, the structures that govern discovery, invention and ingenuity (Chapter 10); third, the public, private and collective processes that engage in gestating new ideas (Chapter 11); and, finally, the distributed governing networks underlying the production and marketing of new products (Chapter 12).

Finally, in Part V, I offer some observations on the implications of using deep governing systems and an assessment of the strengths and weaknesses of our tools of analysis.

## NOTES

1. IENICA (n.d.).
2. AgBios (2005).
3. Mason *et al.* (1996); Tacket *et al.* (1998), (2000); Biemelt *et al.* (2003); Thanavala *et al.* (2005); Maloney *et al.* (2005).
4. CBI (2004).
5. Mason *et al.* (1996).
6. American Ag-Tec (1998).
7. Simon, (2000).
8. Losey *et al.* (1999).
9. Ewens and Pusztai (1999).
10. Smyth *et al.* (2004).

# PART II

# Grounding the analysis

# 2. Transformative technology

> Technology is society, and society cannot be understood or represented without its technological tools.
>
> **Manuel Castells**[1]

## INTRODUCTION

Discussions about technological transformations in society have been all over the map in recent years. In order to focus this analysis, it is necessary to clarify what transformative technological change entails. The best way to start is with the example of the so-called 'information revolution', which involves an array of inventions and innovations related to information, communications and telecommunications. Among other technologies, the 'revolution' involves the silicon chip, the facsimile machine, fiber optics, mobile phones, digital technologies, personal digital entertainment systems and the Internet, including all its attendant technologies and services. There has been a cacophony of debate in the recent past about this revolution and its scope, scale and impact on the global economy and society. There has been enough written about the topic to fill a modest-sized library (a recent search of the British Library of Economics and Political Science truncated the search after finding more than 10 000 directly related accessions and more than 2660 accessions dated in the 2000–5 period) and to clog the less formal pathways for debate (a Google search in late 2005 revealed 14.5 million results and a search for 'information revolution' revealed approximately 29.6 million pages).

In spite of all of this analysis and debate, what we are really talking about remains somewhat fuzzy. Many of the discussions focus on the marvels of the technology itself, others focus on its pathways of development, adaptation and adoption, and still others address the impacts of the technology on the economy and society. Although many of these analyses and commentaries offer useful insights into the progress of the technology, it is often hard to discern the essence and impact of the technology from the array of views expressed.

In order to sort out the underlying features from the noise, it is useful to compare and contrast two contemporary technologies that are part of the broader revolution: the Internet, a publicly accessible worldwide system of interconnected computer networks that transmit data by data packaging using standardized protocols; and the iPod, a digital entertainment system first produced by Apple Corporation in 2001. Each is a part of the information revolution and exhibits critical features of technological change.

The Internet emerged in the early 1970s as part of an Advanced Research Projects Agency research project on 'internetworking'.[2] Although there was significant effort to develop computer-to-computer communications in order to facilitate the exchange of 'packets' of data, it became clear that technical barriers between different systems might result in an array of parallel networks that could not communicate. In response, in 1973, two researchers (Robert Kahn at ARPA and Vinton Cerf at Stanford) collaborated on the design of an architecture that would allow these separately emerging packet networks to exchange information. The resulting Transmission Control Protocol and the Internet Protocol, known as TCP/IP, provided a common platform for exchanging data. Although the evolution of the system was far from straightforward, the now nearly global system enables the asynchronous exchange of a wide range of data from multiple sources and involving multiple media – including electronic mail, journal articles, books, government documents, music, photos and movies – and in two-way real time, synchronous communications, including, among others, on-line chat through discussion forums, audio and video conferencing and increasingly sophisticated multi-player role playing gaming.

This new technology has made it increasingly easy and valuable to enter and engage in the internetworked world. Children are finding they can find more information about more things faster and easier than in the past and can connect with a wider array of people with common interests (be they sports, hobbies, current events or on-line gaming). Students are no longer limited by space – they can draw primary and secondary data from increasingly varied and competitive sources, ranging from the formal and established (such as government websites, on-line access or publication of peer reviewed journals and electronic editions of the quality press) to the traditionally marginalized (including remote and undercapitalized governments in developing countries, public interest groups and the alternate press). Workers are no longer bound to their desks and their environments – they can draw upon the technology to acquire data and information remotely, engage in telephone and video conferencing, transfer data electronically or engage in commerce almost anywhere in the world in a matter of seconds. Consumers can search and buy from an array of on-line and remote suppliers. Citizens can engage in public discourse, either within a narrow, informal community or in regional, national or global initiatives led by governments, intergovernmental agencies or public interest groups. Miniwatts International, a marketing consultancy, estimates that in 2005 about 938 million people used the Internet, equal to about 14.6 per cent of the world's population; this was up from only about 16 million users worldwide in 1995. Usage was highest in North America, where 223 million or 68 per cent of the population was on-line. Although usage was relatively lower elsewhere, it was significant in almost all regions of the world: 269 million or 37 per cent of Europeans were active; 68 million or 12.5 per cent of Latin Americans were on-line; 324 million or 9 per cent of Asians used the Internet; and less than 2 per cent but still more than 16 million Africans were active.[3] The supply of Internet-based content is equally impressive. Internet Systems Consortium, Inc., reported that in July 2005 it found 353 million active IP domains (calculated by pinging the domains and counting the response after deducting duplicate sites), up from only 1.3 million in January 1993.[4]

In contrast, the iPod, though a significant technological and marketing achievement, has a much more prosaic history and impact. The iPod is a brand of portable digital MP3 audio player designed and marketed by Apple Computer. Devices in the iPod family offer a Mac or Windows-based operating system, a simple user interface designed around a central scroll wheel, store media on a built-in hard drive, can serve as an external data storage device when connected to a computer and offer an array of complementary functions, such as personal digital assistance, electronic games and photo storage. Although the iPod was conceived by a researcher outside Apple Computer, the company hired the inventor as an independent contractor to lead Apple's Industrial Design Group to bring the product to market. The first iPod was released on 23 October 2001. Since then four more generations have been developed and released and a number of variations on the technology introduced. Similarly, a number of other companies have developed complementary technologies to work with and enhance the value of the iPod. Accessories include memory-card readers, FM tuners (such as iFM and iTrip by Griffin Technology), voice recording modules, speaker systems (such as made by Bose), in-car audio interfaces (such as in new BMW cars), games, and an array of accessories, style conscious cases and tattoos. *Fortune* magazine reported on 27 June 2005 that Apple has sold over 15 million iPods, including 5.3 million in the first quarter of that year. The iPod currently dominates the digital audio player market, with an estimated 90 per cent of the US market for hard-drive-based players and over 70 per cent of the market for all types of players.

Although both technologies have changed parts of our lives, they are quantitatively and qualitatively different. The Internet is truly a global technology, engaging a wide range of suppliers (more than 353 million in 2005) and consumers (934 million users) and provides a platform for new ways to communicate, work, play, govern and engage the world around us. The Internet is becoming ubiquitous. In contrast, the iPod, arguably a technological and marketing marvel that adds value to the lives of more than 15 million users worldwide, is fundamentally a personal and, for the most part, end-use technology – it is not an input to other production systems, but rather offers specific utility to its owner and user. Both exemplify technological change, but only the Internet offers the opportunity to transform the economy and society.

## HISTORICAL TRANSFORMATIVE CHANGES

The information revolution offers insight into the challenge of transformative technological change, but it is only one of the many examples that are studied. There is no definitive root technology. Typical innovations studies have focused on technological ideas (including fire, the steel ax, the horse and boiling water), civic organization, agriculture, education, medicine and drugs, new communications and other information technologies and new consumer products and on a range of social units (including tribes, villages, individual farmers, firms, individuals and organizations). Perhaps the oldest and most pervasive technology is fire, which has enabled us to cultivate, capture and digest new sources of protein and fiber, to extend our climatic and geographic reach, to manufacture new and more sophisticated materials and tools, to illuminate, to communicate and, ultimately, to captivate our souls. No one truly knows when fire was 'domesticated', but it was most certainly one of the first elements that came under the dominion of man and arguably was a precondition to civilization.

An early set of innovations involved the fashioning of tools, objects and other goods to enhance our economic, social and cultural lives. Well before recorded history there is evidence that man adapted and adopted natural materials to be tools and goods. Anthropologists and archaeologists have generally broken the evolution of materials into three periods, which provide insights and clues into the influence of new technologies on the economy, politics, society and culture of the period. The Stone Age (1 million BC to 3300 BC) encompassed the period when man moved beyond using plants, animals and other naturally occurring materials as tools, to a time when individuals and communities fashioned stones into tools (such as spear or arrow tips and clubs) and perfected the art of pottery. The Bronze Age (3300 BC to 1200 BC) emerged as man developed and advanced the techniques of smelting natural outcroppings of copper with tin, zinc and other alloys to cast bronze, and thereby fashion more resilient tools, weapons, armor, household goods and ornaments. As copper has a relatively low melting point (at least compared to iron and steel), bronze was feasible to make with relatively simple furnaces and was generally stronger than the iron then available. The Iron Age, precipitated by a shortage of tin in the Mediterranean area around 1200 BC, led to the gradual decline in the use of bronze, especially in the operation of the economy and in warfare. Each of the changes in the dominant materials of the day led to economic, political, social and cultural changes in society.

Modern studies of materials have tended to focus on the efforts during the early years of the industrial revolution to perfect the smelting of steel. Stanford economist Nathan Rosenberg, for example, undertook extensive study of the evolution of steel processes.[5] Steel was traditionally manufactured by heating a mixture of charcoal and wrought iron for periods of up to a week. This process was gradually replaced in the 1740s with the introduction of Benjamin Huntsman's crucible steelmaking technique, which reduced the firing time to about three hours but still required significant quantities of expensive coke. The Bessemer process, patented by Henry Bessemer in 1855, involved removing impurities from the iron by blowing air through the molten iron, which also raised and sustained the temperature of the material. The result was a purer form of steel that was produced in as little as a half hour and used only a small amount of coke, thereby reducing the cost per ton from about £40 to as low as £7.[6] Sir Carl Siemens subsequently improved the open-hearth furnace (1865) and Percy Gilchrist and his cousin Sidney Thomas modified the Bessemer process in 1878 (phosphorous and sulfur, which make steel brittle, were removed by lining the converter with limestone and adding lime to the mix), helping to extend the technology to France and Germany, where high phosphorous ores were used. Worldwide steel production rose from only about 22 thousand tonnes in 1867 to 28 million tonnes by 1900.[7] The diffusion of high quality, lower-priced steel made a wide range of industrial, military and consumer products now possible, and the corresponding industrial revolution caused wide ranging, fundamental changes in the economic, social and military relations of the day.

Materials science advanced significantly in the twentieth century, with a wide range of metals and industrial materials being developed, improved, adapted and adopted. As the century ended, much of the focus was on the area of silicon and new ceramic materials, which involve a fusion of metallurgy, chemistry, physics and engineering.[8] Silicon, a tetravalent metalloid which is less reactive than its chemical analog carbon, makes up about one quarter of the Earth's crust. Silicon has a number of traditional roles as an ingredient in many other materials (including enamel, steel, bronze, glass, abrasives and bricks) and in some newer, advanced technologies (such as semiconductors, photonics, fiber optics, medical implants, LCDs and solar power panels). Perhaps the greatest single transformative use of silicon to date has been its 'doping' with boron, gallium, phosphorus or arsenic, to produce silicon for use in transistors, solar cells, rectifiers and other electronic solid-state devices. In short, silicon is the foundation of the information revolution. Others have suggested that silicon, in all its forms, is most important because of its ability to recombine with other materials and its role in an increasing array of downstream innovations, and its pervasiveness in many applications.[9]

Inventions and innovations related to materials were often co-evolving with new means and structures for the harnessing, generation or transmission of power. Although the domestication and subsequent breeding of animals (including dogs, horses, camels and donkeys) marked our early attempts to enhance man's power, most of our analysis has been focused on a number of major innovations of the industrial revolution: the waterwheel; steam engines; electricity and electric motors; power delivery systems; the internal combustion engine; and related transportation innovations (rail, steam boat, automobiles, container ships and airplanes).[10] Most of the studies have highlighted the wide range of complementary uses of each of these innovations, which correspondingly sparked long-term efforts to exploit the new opportunities. The steam engine, for example, is credited with powering the industrial machinery and related transportation systems (rail and steamship) that precipitated and sustained the industrial revolution of 1760–1830. The greater application of steam power facilitated the replacement of labor with machinery, displacing labor to new industries, or, more frequently, to populate colonial outposts in the burgeoning European empires.

Steam power also transformed the printing sector, enabling a massive expansion of newspaper and popular book publishing, which some assert reinforced rising literacy and created demand for mass political participation. Similarly, the development of electricity and electric motors (in combination with the improvements in steel production) is credited with providing the basic impetus for a second industrial revolution (beginning about 1870). Among other developments, the electric motor is credited with enabling reorganization of industrial production into assembly lines, most spectacularly developed by Henry Ford. Previously, manufacturing and assembly were organized to optimize the use of power (high-energy activities were closest to the source of power and vice versa); with electrical engines, it became possible to distribute power to all parts of a production plant, enabling rationalization of the assembly process. The attendant greater specialization of labor led to sharply higher productivity but greater alienation of workers. Some attribute the rise of trade unions and socialism to the consequences of the two industrial revolutions.

The information, communications and telecommunications (ICT) revolution has more nebulous and correspondingly indeterminate roots. One could date the revolution to the adaptation of the Chinese invention of movable print or to its subsequent introduction into Europe by Johannes Gutenberg in Germany around 1450 and adoption in England by William Caxton, who in 1474 published the first book in English. Early printers had a great influence on the standardization of the written word, as they chose common spellings and grammar. Subsequently, the spread of literacy and the rise of the popular press in the first industrial revolution effectively opened up society to new debates and modes of work.

Finally, the adoption of electrical transmission of information, via telegraphs, began in the 1830s. Semaphore systems had been used for the previous 40 years in Europe but were slow and expensive and thus inappropriate for transmission of all but the most vital security and commercial information. The electrical telegraph was independently developed late in the 1830s in the UK by Sir Charles Wheatstone and Sir William Fothergill Cooke and in the US by Samuel Morse. Morse enhanced the technology with his new Morse code signaling alphabet. Transatlantic cables were successfully laid in 1866. A range of improvements in the telegraph and transmission lines over the following years combined to reduce the cost of sending a telegraph message by more than 95 per cent from the earlier semaphore system. Simultaneously, a number of private inventors in the US and Europe were working on a machine to transmit the spoken word – Alexander Graham Bell in 1875 ultimately developed and named the invention the 'telephone'.

By the end of the century the focus had shifted to wireless technologies. Guglielmo Marconi sent and received his first radio signal in Italy in 1895 and by 1902 he had completed a successful transmission across the Atlantic. The invention of the teletypewriter in the early 1900s signaled the end of hand transmission and the beginning of mass transmissions of data. While the telex data transmission system continued to evolve (with new subsystems and coding languages emerging for special purposes), new technologies were gaining greater acceptance. The telephone system expanded, to the point where by the middle of the century most people in Europe and North America had access to, if not subscribed to, a phone service. This set of developments sharply reduced the barriers of distance and raised awareness of more alternatives for how we work, live and govern ourselves.

The next wave of the information and communications revolution resulted from the invention of the computer. Although many credit Charles Babbage with conceptualizing and designing the first fully programmable computer as early as 1837, it was never constructed. Machines that could undertake discrete computations emerged throughout the early part of the twentieth century, becoming through the second half of the century increasingly sophisticated, powerful and efficient with the development of new hardware (valves, transistors, integrated circuits and increasingly large memory chips) and new software (new programming languages, algorithms, operating systems and programs), ultimately interconnecting through the Internet.[11] Some have argued that the new information technologies have fundamentally altered the world, integrating people and institutions into new, instrumental, global networks.[12] The ITC revolution has thereby challenged the concept of space (from the local to the global), transformed corporate structure, changed the nature of work and created a culture of 'virtual' relationships.

One of the most profound transformations we have undertaken in the past 6000 years is our ability to mold nature to our needs. The advent of civilization is inextricably linked with the introduction of agriculture, as a process of cultivating and managing nature.[13] Studies of the foundational innovations in agriculture – the domestication, cultivation and selection of plants and the domestication and selective breeding of animals for power and food – are the domain of anthropologists and archaeologists. Two historical revolutionary periods have been extensively mined for lessons about transformative processes. First, the British agricultural revolution of the eighteenth century, which involved enclosure, mechanization, crop rotation and selected breeding, is credited with creating the conditions for the follow-on industrial revolution. Second, the 'green' revolution, dating from the mid-1940s, involved planned investment by governments, aid agencies and philanthropic groups to develop a series of dwarf wheat and semi-dwarf rice varieties that could produce high yields in semi-tropical climates with the application of high doses of fertilizer and extensive irrigation. The corresponding adaptation and adoption of the resulting hybrid technologies and high-yielding germplasm caused yields to rise three-fold and enabled traditional food-deficient regions to expand production, some to be self-sufficient and a handful to become food exporters (such as India).[14]

Meanwhile, there has been a parallel transformation in the medical and pharmaceutical industry, as traditional knowledge about our bodies has been gradually replaced by a greater understanding of our physical, genetic and metabolic systems. This has translated into a pharmacological explosion, as people everywhere are seeking cures or remedies for all manner of biotic and abiotic factors in our life, ranging from deadly cancers and hereditary genetic diseases, to annoying aches and pains and to cosmetic enhancements. This effort has engaged and drawn on a wide range of disciplines and actors, including indigenous communities who sustain traditional knowledge of herbal or inorganic remedies, modern biochemistry, molecular and cell biology, immunology, neurobiology, advanced scientific instrumentation, complex and powerful computer aided research tools and, perhaps most importantly, a global network of medical practitioners, scholars and researchers in the biomedical sector. This sector has grown from only a minor portion of the economy, to the point that it is now the largest and arguably the fastest growing economic activity in North America and Europe. Beyond the commercial impact (the health and personal care industry was the single largest contributor to US GDP in recent years) and services the sector offers (which have been significant in many areas, such as the eradication of smallpox and the treatments of cancers), the industry has for some led to a dramatic social transformation, with the medicalization of human life.[15]

This cursory review of studies of historical major transformative technological changes offers some insights into how one might characterize transformative change. In most cases these major changes involve upstream inventions that open up a wide array of new production, consumption, political and cultural opportunities; in contrast, non-transformative changes would tend to offer relatively self-limiting end-use solutions. Transformative change would appear to involve an array of developments in concepts, technologies, products and organizational structures, spanning decades if not centuries; non-transformative changes are frequently short-lived. The impetus for change would appear to emerge when two or more knowledge, technology, commercial, political or social systems converge. Although transformations may or may not have significant social or cultural roots, they almost by definition have social impact. Finally, it would appear that as they get going, they tend to precipitate debate, discourse and conflict.

## DEFINITIONS OF TECHNOLOGICAL CHANGE

There has been a deluge of catchphrases, terminology and buzz words used to describe the nature of technological change. Economists, political scientists, sociologists, anthropologists, geographers, marketers, managers, psychologists, industrial engineers, public administrators, pundits and charlatans have all engaged in the debate about what precipitates change and the challenges of exploiting, managing, redirecting or stopping the attendant consequences of change. The terminology ranges from the histrionic ('revolutions') to the bland and technocratic ('induced technological change'). Although the literature on technological change is far from perfect, many of its contributions highlight potentially critical factors.

First, let's start with the meaning of 'technology'. The *Oxford English Dictionary* (OED) defines it as 'a discourse or treatise on an art or arts; the scientific study of the practical or industrial arts; practical arts collectively; a particular practical or industrial art'. As such, it encompasses the applied, useful tools and methods man has made that may or may not be grounded on understanding of the underlying laws of nature (commonly labeled science). Harvey Brooks from the John F. Kennedy School of Government isolated and codified the relationship between science and technology, arguing that science contributes to technology by: contributing new technological ideas; offering tools and techniques for design and evaluation; providing instrumentation, laboratory techniques and analytical methods; creating high-quality personnel; creating new knowledge for assessment of technology; and offering a knowledge base for new research, development and commercialization strategies.[16] Conversely, technology contributes to science by providing a source of questions and a prolific source of new instrumentation. Economic historian Joel Mokyr argues that 'science is controlled observation of the world around us [whereas] economic production is manipulation of observable reality for the special purpose of providing commodities and services desired by human beings'.[17] Some believe science and technology can be studied separately, but Rosenberg suggests that science is inextricably linked in a dialectic way with technological progress: 'under modern industrial conditions . . . technology shapes science in the most powerful of ways: it plays a major role in determining the research agenda of science'.[18]

When the focus shifts to change in technologies, there is often confusion between the concepts of innovation, invention and technological change. Innovation is generally viewed as the adoption of a new useful or significantly improved idea, good, service, process or practice (Chapter 3 will examine the literature and concepts further). University of Minnesota economist Vernon Ruttan suggests invention, in contrast, requires generating 'new things' through an 'act of insight' (this fits well with the patent system, which defines invention as something that is new, useful and non-obvious).[19] Finally, technological change involves a shift in the specific production system. Hence, Austrian economist Joseph Schumpeter concludes 'we simply define innovation as the setting up of a new production function and "technical change" is a shorthand expression for any kind of shift in the production function'.[20]

There are two main ways scholars and others have described and characterized technological change. Some define the change in terms of the scope of the impact of change (such as revolutionary/evolutionary, radical/incremental, drastic/incremental, disruptive/sustaining, breakthrough/iterative or pervasive/ephemeral) at the same time as others focus on the scale and nature of the technology involved (such as hybrid, macroinvention/microinvention, major/minor, general purpose, enabling or transformative). Each of the characterizations offers some insight into the process.

## Scope of Change

Quite a number of scholars and popularizers have termed major shifts in the dominant technology as 'revolutionary'.[21] The term 'revolution' is used either to characterize major transformative events that destabilized extended ages (such as the stone, bronze, iron or steam ages) or characterized the sectoral focus of a transformation (including the agricultural, industrial, information, green and, some assert, biological, molecular or genomic revolutions). Given that revolution is defined as 'an instance of great change or alteration in affairs or in some particular thing', this implies that a technological revolution must be relatively sudden and impose a major change. Although revolutions often begin with bursts of 'optimistic fervor' they usually span a long period.[22]

Mokyr undertook an historical odyssey through an array of historical 'revolutions', generally concluding that past revolutions, regardless of how rapid the technological change (each involved some technological watershed), were slow to show up in the data on economic growth, real wages and living standards; it generally took more than a century for substantive change to be obvious.[23] Economic historian David Landes thus argues for the cautious use of the term – because in most cases statistics are incapable of validating with any accuracy often subtle (and at times dramatic) changes in the economy – noting that some economists have, for this very reason, suggested we abandon the almost iconic term 'industrial revolution'.[24]

A group of six faculty members at Rensselaer Polytechnic School of Management and Technology, led by Richard Leifer, coined the term 'radical innovation' to typify changes that transform the relationships between customers and suppliers, restructure marketplace economics, displace current products, and often create entirely new product categories.[25] Radical innovation thereby provides a platform for the long-term growth that corporate leaders desperately seek. In contrast, smaller, less significant innovations are deemed to be incremental. Economist Elhanan Helpman similarly uses drastic technologies to represent those events that involve a discontinuity, in the sense that an old process or a material is replaced with a new one. Incremental innovations, then, are induced, secondary changes.[26]

Alternatively, Harvard Business School professor Clayton Christensen defined a disruptive technology as a new technological innovation, product, or service that eventually overturns the existing dominant technology in the market, despite the fact that the disruptive technology is both radically different from the leading technology and often initially performs worse than the dominant technology according to existing measures of performance.[27] Sustaining technology then refers to the successive incremental improvements to performance that market incumbents incorporate into their existing product. Christensen subsequently replaced the term with 'disruptive innovation' to reflect the fact that few technologies are intrinsically disruptive or sustaining in character; rather, it is the innovative process that creates the disruptive impact.

Others have offered an array of terms to characterize the scale and scope of change, such as breakthrough,[28] metamorphic[29] and pervasive[30], but the components of their definitions and concepts vary little from the others offered.

## Nature of Change

There has also been a range of categorizations of technology that relate to the nature of the technology itself, rather than the scope of its impact. Joel Mokyr surveyed a variety of these terms.[31] He notes that some talk of 'hybrid' inventions, where different components and epistemologies are blended and recombined in technological hybrids.[32] He also offers the concept of macroinventions and microinventions. A macroinvention 'is one that cannot be considered an improvement or an elaboration of existing techniques even though it, too, must rely on an existing epistemic base. The main characteristic of a macroinvention is an observable epistemological discontinuity from what was possible before.'[33] Economist Zvi Griliches suggested that many breakthrough discoveries or macroinventions actually involve 'invention of a method of inventing', where a major new method of invention leads to a succession of incremental inventions (in his case, he looked at the sustained impact of a method of hybridizing corn).[34] To succeed, macroinventions require 'an act of technological rebellion and heresy', namely 'stepping outside accepted practice and design'.[35] He suggests 'one useful way to think about the economic history of technological progress is to think of it in terms of evolutionary trajectories that begin through a sudden novelty or macroinvention, which then are continuously improved and refined through a multiple of microinventions'.[36]

Rosenberg offers the concept of a 'major' innovation 'that provides a framework for a large number of subsequent innovations, each of which is dependent upon, or complementary to, the original one'.[37] He points to the electrical power plant, the transistor and the computer as quintessential examples. Ruttan also uses the concept of a 'major or strategic invention', only broadening the concept to represent the cumulative synthesis of many individual inventions.[38] Economists Richard Lipsey and Clifford Bekar suggested the concept of an 'enabling technology', which defined by its extensive range of use and complementarities is very similar to Rosenberg's.[39]

A somewhat more detailed and arguably wider concept is the 'general purpose technology' (GPT) offered by economists Timothy Bresnahan and Manuel Trajtenberg.[40] A GPT initially has much scope for improvement and eventually comes to be widely used in many uses by many users. GPTs by definition exhibit Hicksian and technological complementarities, where Hicksian complementarities involve lower factor prices driving substitutions and technological complementarities occur whenever one technological change requires a redesign or reorganization of other production systems.[41] Bresnahan and Trajtenberg suggest the category of GPTs is best viewed as a natural extension of the concept of 'enabling technologies' because they open up new opportunities rather than offer complete, final solutions. They posit that all technologies exhibit a tree-like structure, with a few foundational, prime mover technologies and many derivative and incremental technologies. General purpose technologies often start by addressing only a single or a few purposes but increase in sophistication and efficiency as they diffuse throughout the economy. They cite a variety of examples of GPTs, such as the electric motor, microelectronics, ICT (including writing, printing and the silicon revolution), materials (bronze, made-to-order materials), power delivery systems (water wheel, steam, electricity, internal combustion engine) and transportation (rail, motor vehicle) that involve 'innovational complementarities'. The prevailing view is that GPTs have three key features: they are pervasive (used widely in multiple sectors); they are dynamic (open up opportunities); and they offer two types of complementarities (between GPT and the application sectors and across the application sectors). Richard Lipsey has suggested that the open-ended and indeterminate nature of GPTs may mean that market forces are unable to pick or nurture the optimal number and range of GPTs, which could lead to a market failure that could justify public intervention.[42] Ultimately, all agree that GPTs generate economic, social and political changes.

British sociologist Perri 6 proposes the concept of 'transformative technologies' which are 'involved in large-scale, discontinuous change to societies'.[43] Transformative technologies (such as printing, the steam engine, electricity and atomic energy) are so fundamental because they have many varied purposes, have applications in a large part of society and the economy, exhibit significant scope for improvement initially, have strong complementarities with other technologies and have a long-term effect on values, power structures and ideas.[44] Transformative changes are distinct from iterative ones, in that they: draw on cutting-edge basic as well as applied science; represent step changes in the scale and in some cases the direction of development of human capabilities (they are not simply iterative, incremental enhancements of existing capabilities); have consequences distributed widely over many areas of life; are of sufficiently high profile to attract the attention, interest and risk perception of social movements, citizens, politicians and regulators; precipitate public debate and the attention of futurists, journalists, ethicists and technologists; and are within a few years or at most a decade of realization.[45]

Although there is a range of views about the nature of transformative technology, looking at the definitions in their entirety offers a vibrant picture. A few elements stand out. Transformative technologies involve disjointed, step adjustments in our productive and institutional capacity, displacing, destabilizing or overturning precursor systems. In technical terms, they often are driven by new epistemologies, they offer significant complementarities (within and across sectors), and they tend to involve convergent, recombined or hybrid technologies (one might actually label them convergent technologies). Furthermore, although they often emerge wrapped in an optimistic fervor, they are hard to anticipate because they infiltrate and influence multiple sectors, markets and domains over a long and variable timeline – in spite of the rhetoric, none of the identified technologies overwhelmed society quickly. This makes it difficult to manage or govern them through established markets or authorities. The nebulous nature of transformative technology creates uncertainty, which tends to generate debate and controversy.

## TRANSFORMATIVE TECHNOLOGIES IN THE TWENTY-FIRST CENTURY

If the root of a transformative technology is that it reflects the convergence of one or more technology domains, then there are a wide range of possible transformations in the current century. Perry 6 identifies nine different transformative technologies that he believes will profoundly influence society in the twenty-first century: network engineering; nano-engineering; quantum computing; bio-computing; robotics; genetic engineering; neuro-engineering; non-polluting production technologies; and new space transport technologies.[46] Although each of these could conceivably be a transformative technology, they can be disaggregated to reveal their roots in one of three core knowledge domains – information, biology or materials.

Starting with the three core knowledge domains (information, biology and materials), bimodal or complex recombinations have the potential to create four hybrids. This typology of seven areas of potential transformation and governing challenges offers one way to look at the interconnections between Perry 6's examples (see Figure 2.1).

The information domain in its purest sense involves language and various schemes of codification. Language can be the tongue or idiom of people, such as English, French or German, or it can be other symbolic and conceptual configurations. Economist Kenneth Boulding argued 'mathematics is a tool or a language' and that because 'all language involves abstraction (orderly loss of information), we can never talk about as much as we know. Mathematics is an abstraction from ordinary language, an abstraction which again involves loss of "richness" of information.'[47] Similarly, the American Standard Code for Information Interchange (ASCII) is a language, in that it specifies the correspondence between digital bit patterns and the symbols of written language, thus allowing digital devices to communicate with each other and to process, store, and communicate character-oriented information.[48] A wide range of codification schemes have evolved, including the Dewey decimal system for libraries, the Institute of Scientific Information's codification of academic disciplines and sub-disciplines and the geographic coding of location by latitudes and longitudes that underlie navigation and the US Department of Defense Geographic Positioning System (GPS).

The second pure domain, the preserve of inorganic materials, initially involved working with metals and stone, but has now expanded to include a

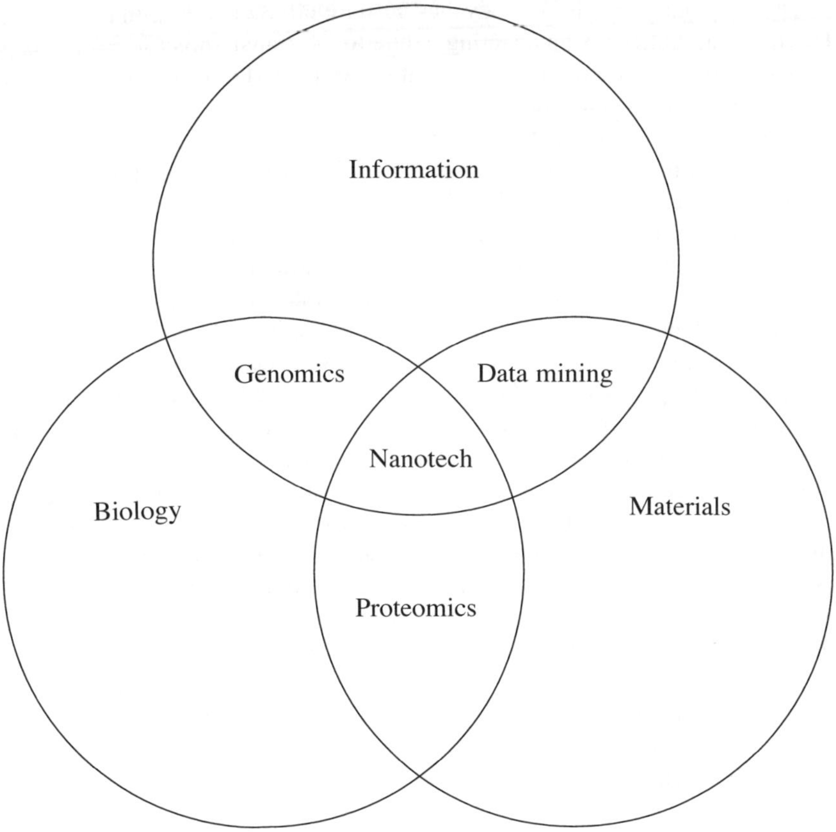


*Figure 2.1 The knowledge domains*

wide range of sub-fields that generally examine properties, structures, performance and processing of both natural and man-made materials. In addition to metallurgy, the study of metals which evolved through the bronze, iron and steel ages, we now have crystallography (the study of the physics of crystals, including their defects and various diffraction techniques), ceramics (involving electronic semiconductors and structural materials), and tribology (study of the wear of materials due to friction and other factors). This domain has revealed both the subtle and often unexpected properties of natural materials as well as assisted engineers and scientists to assemble new materials with specific properties. Most of our industrial products are the direct result of materials science, either in terms of the product itself or as a critical part of the underlying production system.

The third pure domain, biology, has been rapidly expanding in recent years. Though humans have been adapting nature to their own ends for millennia, the modern period of biology is usually dated from Charles Darwin's publication in 1859 of his treatise on biological evolution, the *Origin of Species*. Shortly thereafter, in 1866, Gregor Mendel, a little-known Austrian monk, published a short tract entitled *Experiments on Plant Hybridization* that outlined the results of his experiments with hybridizing peas, and laid the foundation of what is now called the Mendelian laws of inheritance. Even though the concepts were articulated, it took another generation before the laws of inheritance were reduced to common practice and systematically applied to breeding of new plants or animals. The next big breakthrough came in the 1950s, when a group of scientists sought to uncover the nature of inheritance. Although James Watson and Francis Crick are credited with the discovery, the effort actually involved a wide range of scientists from the materials and biological disciplines. They ultimately found that the reproducing parts of living matter, called chromosomes, are composed of DNA, which is a double-helix, chemically linked chain of nucleotides, each of which consists of a sugar, a phosphate and a pair of one of four bases: adenine (A), thymine (T), cytosine (C), and guanine (G). These bond in pairs, A with T and C with G. Sequences of these base pairs form genes, with promoter segments at each end that encode to turn on or turn off the gene. The genes are encoded to produce proteins, which because of their particular shapes and sizes have different functions.[49] This discovery, the foundation of our modern understanding of living matter, was not directly useful. It wasn't until 1973 that Herbert Boyer and Stanley Cohen combined efforts to invent a method of cloning genetically engineered molecules in foreign cells. Boyer's lab at UC San Francisco isolated an enzyme that could cut DNA into precise and 'cohesive' segments that would carry the code for a pre-determined protein and could also be attached to other strands of DNA. Cohen, at Stanford University, had developed a method to introduce antibiotic-carrying plasmids into certain bacteria, as well as a method of isolating and cloning genes carried by the plasmids. The result was a tool that would allow scientists to 'cut and paste' DNA into a genome.[50] The first major commercial application of this technology was the development of human insulin, which was commercialized by Genentech (Boyer was a founder of that firm). These tools have since been used to develop a handful of transgenic plants (herbicide tolerant soybeans, corn, and canola, insect resistant corn and cotton and viral resistant potatoes), which have been planted over the past 10 years on more than 200 million acres in 17 countries, located on six continents,[51] to express industrial and pharmaceutical proteins into plants (such as laurate and hirudin in canola), to clone animals (including Dolly the sheep, Prometea the horse, Ralph the rat, Dewey the deer, Rainbow and CC the cats, Noel the pig, Noah the gaur, Cumulina the mouse and a host of rabbits), to genetically modify a few animals (such as Harvard's cancer research Oncomouse and Nexia's goat that produces spider silk) and a host of new drugs (most notably various AIDS drugs). These developments have generated an unprecedented debate about how far society wants to go in modifying nature.

Beyond the pure domains, hybrids are possible. The definition of transformative technology would suggest these hybrids are likely to be the source of the most dynamic changes. Once one leaves the safe, well-delimited confines of any well-established discipline, it becomes difficult to determine where influence lies. The following discussion is not intended to be a dogmatic formulation of causes and effects, but rather an illustration of how hybrids have led to transformative technologies.

The interface between the information and materials worlds undoubtedly led to the information revolution of the past generation. The increasingly sophisticated uses of materials (especially silicon), combined with evolution of progressively more sophisticated mathematical, computational and information theories and the development of new computer-directed languages, led to the evolutionary development of mainframe computers, minicomputers, microcomputers, networked computers, and a burgeoning number of wireless and other distributed computing structures. Although more evolution is likely, we can generally see the nature and scope of the transformations this hybrid pairing of technologies is likely to bring.[52]

The pairing of the materials and biological worlds has already borne fruit. The discovery of DNA was only possible with the application of materials science methods and techniques. Although the Cambridge-based team of Watson and Crick get most of the credit for discovering the structure of DNA, an American group led by Linus Pauling and a group at King's College London (including Maurice Wilkins and Rosalind Franklin) were both using X-ray diffraction to picture DNA. Thus, we already have one example of a hybrid experiment leading to insight. Currently there is extensive work underway by protein crystallographers who are using X-ray crystallography and nuclear magnetic resonance spectroscopy to determine the structure of different proteins. The goal is to develop new proteins for food, industrial and biomedical uses.

The effort to sequence genomes of plants, animals, microbes and viruses over the past 20 years was triggered by the pairing of new high-speed DNA sequencers, advanced computer programming and new understandings of biology. The most significant effort was the effort to sequence the human genome. What turned out to be a race between a global consortium led by the US National Institutes of Health and the UK Wellcome Trust and a private sector effort by Celera Genomics and its founder, Craig Venter, delivered a rough draft of the human genome in 2000. Along the way, more than 180 other genomes have been sequenced, including bacteria (including *E. coli* and *Haemophilus influenzae*), the fruit fly, mouse, rat, rice, *Arabidopsis thaliana* and puffer fish. Knowing a sequence is only the beginning of the voyage of discovery. We also need to know how the genome segments into genes and what those genes encode for. Although the trial and error method of knocking out a gene and observing the results can work, it is slow and expensive. There are a variety of other attempts to find target genes. For example, a research team from the Mayo Clinic College of Medicine in Rochester, Minnesota, is working on a new method of pairing historical herbal texts and modern medical databases to identify candidate therapeutic proteins in plants.[53] Throughout history physicians, expatriates and explorers have documented the uses of plants by indigenous peoples. Now, the combination of electronic scanning, optical character recognition and sophisticated data mining systems – enabling the comparison of phenotypical, practical and social aspects of plants with pharmacological databases – offers the promise that scientists may be able to identify novel pharmacotherapeutic bioactive compounds that could have value in industry and medicine.

The ultimate destination of many researchers is the convergence of information, biological and materials sciences in what is increasingly called nanotechnology. A nanometer is one billionth of a meter, which is approximately equal to the length of 10 hydrogen atoms. The field is premised on the knowledge that atoms are the fundamental building blocks of all matter, both organic and inorganic and that the atomic structure of all matter ultimately determines it effect. As materials subdivide, surface-area-to-volume increases, causing compounds to perform differently (including changes in color, magnetism, reactivity and conductivity). Though nanotechnology probably has most in common with materials science, the field involves researchers interested in almost all aspects of organic and inorganic compounds. Researchers are seeking ways to exploit their ability to understand the atomic structures of things in order to restructure compounds to affect how they bond, crystallize, fold, form layers or otherwise assemble from their components. Researchers thought of working at this level as early as the late 1950s but actual practical work only began when IBM invented the scanning tunneling microscope which enables researchers to examine the physical structure at the nanoscale.

Lux Research, a consultancy based in New York, defines the nanotechnology value chain as involving four differentiable components.[54] Nanomaterials are minimally processed engineered structures with a dimension of less than 100 nanometers that exhibit size-dependent properties, such as nanoscale titanium dioxide, nanotubes, quantum dots and dendrimers. Nanointermediates incorporate nanomaterials or have been constructed with nanoscale features (some have been used, for example, to convert heavy oils into usable fuels). Nano-enabled end products involve a composite of nanomaterials and intermediates, such as GM's body side moldings made from a nanocomposite material and 3M's Adper Single Bond Plus dental adhesive. Finally, nanotools can be used to visualize, manipulate, and model matter at the nanoscale, such as Veeco's atomic force microscopes and Accelrys' molecular modeling software.

Nanotechnology, at least partly because of its unfathomable promise, is the darling of governments everywhere. Lux Research estimates that governments, companies and venture capitalists combined invested more than US$8.6 billion in nanoscience and technology in 2004.[55] The US federal government was estimated to be investing more than $1.6 billion in 2004, more than twice what it spent on the Human Genome Project at its peak. The main end-users of the technology in 2005 were Air Products and Chemicals, BASF, EI Du Pont de Nemours and Company, General Electric, General Motors, Hewlett-Packard, International Business Machines, 3M and NEC, all chemical, manufacturing or advanced instrumentation companies.

Although the nature of the investment is relatively easily identified, the scale and scope of the market for nanotechnology is more problematic. The applications are touted to be almost limitless. Early marketed applications include antimicrobial surgical bandages, odor-proofed cotton fabrics, stronger tennis rackets, coatings for hulls of boats, sunscreen, and parts of cars and refrigerators. Proposed uses include zero-effluent manufacturing, improved solar cells and a host of industrial uses, including mini biological factories. What is not clear is how much this technology is generating in terms of revenues. Lux has surveyed the prognosticators on the potential size of the market and concluded that up to 2005, less than US$250 million in gross sales was realized, but that by 2015, the range of estimates was for gross receipts to rise to between US$500 million and US$2 billion.[56]

One lesson to be learned from past transformative technologies is that it is very difficult to predict if or how hybrid technologies will perform and disseminate into the economy and society. The governing challenges will likely increase as these technologies converge, combine and recombine into new hybrid areas of knowledge and technology.

In the final analysis, the challenge of transformative technological change is not one of developing new science and technology but a problem of organizing complex human relationships. Complex systems require sophisticated analysis. The following chapters will examine the innovation processes related to transformative technology and the governing challenges these pose.

## NOTES

1. Castells (1996).
2. Kahn (1994).
3. http://www.internetworld stats.com/stats.htm

4. http://www.isc.org/
5. Rosenberg (1994).
6. Wikipedia, accessed December 1995.
7. Ibid.
8. Rosenberg (1994: 148).
9. Bresnahan and Trajtenberg (1995).
10. Bresnahan and Trajtenberg (1995); Lipsey (1995); Stewart (1992).
11. Rosenberg (1994); Bresnahan and Trajtenberg (1995); Lipsey (1995).
12. Castells (1996: 22).
13. Diamond (1997).
14. Evenson and Gollin (2003).
15. UBC health economist Robert Evans has argued that 'health' is increasingly viewed as an undiagnosed state of illness.
16. Brooks (1994).
17. Mokyr (2002a).
18. Rosenberg (1994: 16).
19. Ruttan (2003).
20. Schumpeter (1939).
21. For example, Fukyama (2002); Bruland and Mowery (2005); Persaud (2001); Phillips and Khatchatourians (2001); Mokyr (2002a).
22. Persaud (2001: 108).
23. Mokyr (2002a, b).
24. Landes (1999).
25. Leifer *et al.* (2000).
26. Helpman (1998).
27. Christensen (1997).
28. Kalthoff *et al.* (1997).
29. Darby and Zucker (2003).
30. Michalski *et al.* (2001).
31. Mokyr (2002a).
32. Patvitt (2003), talks of technological convergence.
33. Mokyr (2002: 27).
34. Darby and Zucker (2003).
35. Ibid.
36. Ibid.
37. Rosenberg (1994: 15).
38. Ruttan (2005).
39. Lipsey and Bekar (1995).
40. Bresnahan and Trajtenberg (1995).
41. Lipsey *et al.* (cited in Helpman, 1998: 43)
42. Lipsey (1995).
43. 6 (2001: 74).
44. Woodley (2001).
45. 6 (2001).
46. 6 (2001).
47. Boulding (ND: 1–2).
48. Wikipedia.
49. Ridley (1999), offers a short but illuminating layman's explanation of the science in the introduction to his book.
50. Lemelson-MIT (1997).
51. James (various).
52. Castells (1996); Held *et al.* (1999).
53. Buenz *et al.* (2004).
54. Lux (2005).
55. The Economist (Loder 2005).
56. Ibid.

# 3. Innovation

> All innovation begins with creative ideas . . . creativity by individuals and teams is a starting point for innovation; the first is a necessary but not sufficient condition for the second.
>
> **Amabile, *et al.***[1]

## INTRODUCTION

The challenge of getting a new idea into use is greater than many people realize. The ingenuity or creative spark to see a need and seek to fill it (what economist John Maynard Keynes called 'the animal spirits') is perhaps the genesis, but there is a long and often tortuous path to success. The story of the origin and evolution of the edible oilseed canola, sometimes called double-zero rapeseed or oilseed rape, illustrates the nature of innovation.[2]

Canola provides the third largest source of edible oils in the world. The name canola is given to two particular subspecies of rapeseed plants that exhibit well-defined oil and meal properties: self-pollinating *Brassica napus L* and the outcrossing *Brassica rapa L.* Since the Second World War, extensive research and development has transformed rapeseed into canola and positioned it as a premium edible oil crop. In one sense, the canola story does not involve anything that had not been thought of by Mendel, the father of modern plant breeding. The science of selective breeding to enhance input or output traits goes back to his oft-cited experiments with peas. Nevertheless, there were two periods in recent times where evolutionary and at times revolutionary breakthroughs in science and technology allowed a quantum leap in the development of the industry.

The original transformation of rapeseed into canola and more recently the privately directed, research-intensive activity around canola is largely a Canadian story.[3] For thousands of years rapeseed has been used in various markets as a cooking oil, industrial lubricant, animal feed or green manure. Focused research between 1944 and 1978 in universities, at Agriculture Canada (now Agriculture and Agri-Food Canada, AAFC) and in the National Research Council (NRC) labs in Canada sought to improve the agronomic and food attributes of rapeseed. That effort culminated in 1978 with the development of a new variety of rape that had lower amounts of erucic acid (which was linked in some studies to heart problems, especially for those with high-fat diets) and glucosinolates (which impair the feed value of the meal). The new low erucic acid, low glucosinolate (sometimes called 'double zero') standard was trademarked by an industry association in Canada as 'canola' and became the forerunner of virtually all of the varieties used around the world today.

That first transformation was the product of the effort of a relatively narrow and well-delineated group of applied scientists who, beginning in 1944, proactively sought to develop an edible oilseed for Western Canadian cultivation. Previously, breeding new varieties of canola was extremely difficult. Breeders used a trial and error method, testing a wide variety of seeds, selectively crossing and propagating those that appeared to exhibit desired traits. Testing for oil composition was laborious, taking about two pounds of seed and two weeks to undertake. Given the natural heterogeneity in canola, even when a sample tested well, it was not always possible to be sure that the seed remaining in the sample would hold the same traits. The first breakthrough came in 1957 when the NRC Prairie Regional Laboratory acquired a gas liquid chromatography (GLC) unit and perfected the technique of more quickly assessing the oil properties of increasingly smaller samples. The NRC staff ultimately refined the technique to the point where tests took about one seed and about 15 minutes to complete. The NRC then assisted the breeding programs at the Dominion Forage Crops Laboratory of Agriculture Canada and the University of Manitoba to acquire and use GLC units. A team of scientists at the Agriculture Canada center in Saskatoon further refined the GLC technology to the point where they could test half of a canola seed. They then went on to develop a technique of cutting a single seed in half in such a way that the remaining half could germinate and produce a new plant. They proceeded to test thousands of seeds during 1962 and 1963 to find one with low erucic acid and in 1963 succeeded in finding a single seed with no measurable quantities of that undesirable trait. They grew the half left after the test into a sickly plant that ultimately yielded five whole seeds. Those seeds formed the basis for the first low erucic acid *B. rapa* variety released in 1971. The rest of the breeding program to develop canola adopted and used those two innovations – GLC analysis and the half-seed method – forming the foundation for the modern canola industry. By 1978 both of the two subspecies of canola, *B. napus* and *B. rapa*, were converted using these techniques to exhibit low erucic acid and low glucosinolates. These new varieties were tested for safety and efficacy by the small, relatively close-knit breeders' community through the Seeds Act – candidate varieties were assessed on more than 15 different traits in comparison to prevailing varieties and only clearly superior varieties were allowed to enter the market. Even though the new double-zero varieties yielded 18 per cent less than traditional varieties, the benefits of reducing the two deleterious traits were significant enough that within four years the conversion between technologies was 95 per cent complete. The 'yield drag' was gradually eliminated by incremental varietal improvements over the succeeding decade.

Once genetic engineering technologies were introduced, canola was targeted for further development by private companies. Beginning in the mid-1980s, a number of private research programs identified that canola/rapeseed would be an ideal plant to make tolerant to broad-spectrum herbicides. It exhibited a number of features that made it attractive. First, it has a relatively simple genome and a proven transformation technology (*Agrobacterium tumefaciens*), which made it easier and more predictable to modify. Second, it was a weedy crop and farmers already invested heavily in weed control for the crop. Third, given that canola germplasm was relatively unstable, most producers purchased seed annually. Fourth, there was readily available publicly developed germplasm to use as a base for new varieties.

Four separate efforts began about that time. Three of the research efforts involved the transgenic *Agrobacterium tumefaciens* transformation system. Monsanto began to search for a way of inserting its patented Roundup Ready™ gene in the plant to make a glyphosate resistant variety, and introduced the GT73 variety in 1995. AgrEvo (subsequently acquired by Bayer Crop Science) inserted its LibertyLink™ gene to make a glufosinate ammonium tolerant variety, first cultivated commercially in 1995. Rhone Poulenc began work on a bromoxynil and ioxynil resistant variety about that time but it was only approved for release for the 2000 crop year. Pioneer Hi-Bred, in contrast, adopted a different approach in collaboration with American Cyanamid, using accelerated mutagenesis to introduce imidazolinone (Pursuit™) tolerance into the plant. Each of these programs drew extensively on the intellectual property developed and managed by others.[4] The basic research that underlay that effort was increasingly dispersed. In 1981 only about 22 countries were doing any academic research into canola; Canada was by far the dominant actor. By the mid-1990s more than 40 countries were doing research into canola on a continuing basis and over the intervening period more than 70 countries did some work on canola. The Canadian patent database illustrates the impact of that diffusion of knowledge. Between 1985 and 2000, there were been 634 patents issued for canola-related innovations (45 per cent for process and 55 per cent for product inventions), the vast majority of them to foreign (mostly US-based) research teams and companies. In practical terms, virtually all of the key genetic transformation technologies used by the canola sector were the proprietary tools of non-resident companies.

In 1995 canola became the first genetically modified food crop in the world to be approved for use. The regulatory system had evolved significantly since the introduction of canola in 1978. Instead of simply leaving the review to breeders, the federal government had established an extensive regulatory structure, involving statutorily independent regulators in the Departments of Health, Agriculture and Environment to evaluate and assess the product for its human and animal health impacts and to review its impact on the environment. Since then, there have been ten herbicide tolerant events and two novel oil traits approved and transferred into more than 100 varieties, which collectively captured an estimated 90 per cent of the market share in Canada by 2005. Although the first new varieties delivered approximately 5 per cent lower yields than competing traditional varieties, the yield drag was eliminated in most cases in less than three growing seasons. More recently, there have been confined breeding and regulatory trails in Canada for a range of new varieties modified to be insect and disease resistant and to include specific output traits, such as modified industrial oils, nutraceutical properties and pharmaceutical proteins and enzymes. Along with the changes in the seed sector, there has been a radical shift in the structure of the Canadian production and marketing system. About half of the varieties registered for production in Canada since 1995 exhibit some trait that is managed through some form of identity preservation, with the concomitant evolution of proprietary, close-looped production and marketing chains.

The canola story exhibits a number of features common to most innovations. Changes in both periods were dependent on ingenious and driven proponents of change (be they public sector scientists or multinational companies) but the nature of the innovation process changed during the period. The earlier transformation tended to be supply-push, with a relatively narrow, linear technological approach to developing and commercializing a new product. In spite of the focused effort, the complete cycle from genesis to full technology adoption still took almost 40 years. In recent years, new scientific knowledge has been the driver. As the traditional, slower, trial and error technology-driven approach was supplanted by a science-based system, the innovation process was transformed from a relatively simple supply-push, linear research and development system into an increasingly complex, demand-pulled, dynamic and interactive research and development process with extensive loop-backs, programmed production and intensive knowledge management. One result, at least in this case, is that transgenic technologies have matured and dominated the market in about half of the time of the earlier innovations.

## INNOVATION IN THEORY AND PRACTICE

As with much of the literature about technology and the economy, Joseph Schumpeter is credited with the first economic definition of innovation. He defined innovation as the introduction of a new good or a new quality of an existing good, a new method of production, the opening of a new market and the introduction of a new supply of inputs to a production system or a new organizational structure in an industry.[5]

Two points are clear from his definition. First, he very explicitly separates invention from innovation, arguing that 'innovation is possible without anything we should identify as invention, and invention does not necessarily induce innovation'.[6] Although some might argue that innovation involves introduction of a 'better way of doing things', one should keep in mind that many, if not most, new technologies typically emerge in a relatively primitive condition.[7] It is increasingly becoming accepted that it is irrelevant whether an innovation offers any advantages over the product, technology, market or organization it is replacing; rather, all that matters is whether individuals perceive an advantage.[8] In the canola example above, for instance, the first double-zero varieties generated some health benefits for consumers with high-fat diets (such as in North America and Europe, but not in Japan where heart disease was not a major health factor) but mixed agronomic benefits (reduced glucosinolates made the meal suitable for animal feed but the new double-zero varieties produced about 18 per cent lower yields than the traditional rapeseed varieties they replaced). Sustainable innovation, however, most certainly requires a net gain to society (in our canola example, the yield drag was eliminated quickly, making this a win–win technology). Regardless of these nuances, much of the literature and all of the public policy interest tends to focus on those innovations that bring something new – but especially an invention – into widespread, practical use.

Second, Schumpeter is clearly talking about a process ('introduction') by which something new is initiated or adopted. Ruttan concurs, suggesting the concept of innovation be extended to include the process by which new things emerge in science, technology and art.[9] A variety of other researchers see creativity – defined as the production of novel and useful ideas in any domain – as the basis for innovation and so they define innovation as the successful implementation of creative ideas within an organization.

Much of what people identify as innovation involves planned, systematic effort to add value through research and development.[10] Stanley Metcalfe, professor of political economy and director of Policy Research in Engineering, Science and Technology at Manchester University, suggests that a number of key stages occur with most innovations: investments are driven by innovation possibilities anticipated in the minds of technologists, research managers and innovators; it is uncertain whether the research, development and commercialization inputs will generate any innovative output; and innovation co-produces targeted useable output (which might be 'agenda reducing') as well as enhances the number of possibilities (which is 'agenda enhancing').[11] The Committee for Study of Invention produced a report for the US National Science Foundation that explicitly differentiates invention ('the process of devising and producing by independent investigation, experimentation, and mental activity something that is useful and that was not previously known or existing') from innovation, which they define as 'the complex process of introducing novel ideas into use or practice. Inventions, thus, generate value from their use (innovation) and not from their creation and existence'.[12]

There are two main competing views about the processes of innovation. The economic theory of technological change articulated by Nobel Prize winning economists Kenneth Arrow and Robert Solow generally focuses on the firm as the primary research unit. Other economists, in the footsteps of Schumpeter, have examined the microeconomic incentives and impacts of private research for commercialization. Some economists recently have examined the impact innovative activity at the firm level has on the larger economy, focusing especially on the implications of 'endogenously' generated innovation on macroeconomic growth, trade and industrial location.[13] Even so, they tend to assume that innovation is a discrete event that occurs within firms. Recently a few economists have begun to note that firms are not self-sufficient and often must reach beyond their boundaries to supplement their capacity. Nevertheless, their focus remains steadfastly on innovation that is primarily directed by a single corporate leader. Alternatively, a group of political economists and sociologists with more of an interest in the influence of institutions has developed a range of theoretical 'systems' approaches to innovation. Most hark back to Alfred Marshall and his *Principles of Economics* and look at the role of economies of scale and scope in the local, regional, national or sectoral innovation systems themselves. Once again, however, this approach tends to consider innovation as an activity that is or should be limited to a narrow subset of actors.

The OECD Oslo Manual, which provides guidelines for collecting and interpreting technological innovation data, suggests innovation is perhaps better looked at as a nested activity, with a complex set of factors within firms creating an 'innovation dynamo', which directly relies on an array of 'transfer' human, social and cultural factors (networks and other linkages between firms) that influence information transmission and learning, and a general set of 'framework' conditions and institutions (including education, financial, legal, public, scientific and cultural endowments) which set the range of opportunities for innovation.[14]

There is significant blurring of the lines between activities and actors, leading to difficulty in observing and assessing the efficacy of the differing modeling efforts. Some see innovation as fundamentally an intra-firm event,[15] others see innovation as an industrial event,[16] a community event,[17] a national event[18] or a collective event among a subset of specific institutions (such as the triple helix of industry, government and university articulated by sociologists Henry Etzkowitz and Loet Leydesdorff[19]). Innovation can thus involve more than the narrow innovation systems, but can encompass a wide range of actors who are geographically disbursed around the world.

In this context, there are four somewhat different literatures related to how the innovation processes operate. Neoclassical economists tend to model innovation as an 'induced' event. British economist and Nobel Prize winner Sir John Hicks proposed as a macroeconomic hypothesis that 'a change in the relative prices of the factors of production is itself a spur to invention, and to invention of a particular kind – directed to economizing the use of a factor which has become relatively expensive'. This he called 'induced innovation'.[20] Rosenberg has credited Charles Babbage, the inventor of the analytic computer, with first making the point that an extensive division of labor enables a succession of applied solutions to small problems; only a few 'beautiful combinations' are the work of occasional genius (what others might call drastic, revolutionary or transformative change).[21] Thus, Babbage spent significant energy on careful cost accounting to identify areas of possible profitable innovation. Considerable literature has been produced on this hypothesis, which is often presented in terms of the effects of wage increases as an inducement to labor-saving innovation. The hypothesis has also been applied to viewing increases in energy costs as a motivation for a more rapid improvement in the energy efficiency of goods than would normally occur. Rosenberg cautions, however, that this dominant focus on cost-reducing technological innovation is dangerous, as it is likely that product innovations and qualitative improvements may have been and likely will be the most important long-term contribution to human welfare.[22] He also suggests that our focus on the processes of large firms may contribute to this misdirection of energy.[23] Large firms tend to generate iterative technological change (after all, they have the most to gain from improving the efficiency of production) but smaller firms tend to generate lower cost but more original innovations that may change the quality or nature of the product or the production system. He suggests that social and industrial structures within a country may influence the balance between induced and broader innovations. In the UK strong craft traditions, with an emphasis upon pride in workmanship, individuality and high standards of product quality, support broader innovation whereas the Japanese keiretsu, engaged in oligopolistic competition, operate management systems that encourage full understanding of existing processes via rotating management, thereby encouraging induced changes but stifling category-busting innovation.

An offshoot of the economics of induced technological change is an interest in calibrating and estimating the returns on investments in areas. The returns to technology literature has expanded greatly in recent years.[24] Many of the earlier analyses adopted a very narrow comparative static approach: costs were simply the net present value of the direct expenditures on the generation of the new technology and the revenues were the net present value of any induced changes (positive or negative) in consumer or producer surplus. This approach tended to generate high rates of return on investment.[25] Recently economists Julian Alston, George Norton and Phil Pardey posited a wider concept of innovation, broadening the analysis to include four general phases of innovation where costs may be incurred or benefits realized: the research phase (involving research and development outlays); the gestation phase (when the invention is reduced to practice, reviewed for safety and efficacy and packaged); the adoption phase (when the product enters the market and private and social returns from direct use are realized); and the knowledge stock phase (when knowledge is stored, dispersed and accessed by others).[26]

Beginning in the early 1960s, there was an increased focus in the economics literature on other approaches to the study of technological change. Nobel Prize winner and economist Kenneth Arrow introduced the concept of learning by doing, which began the examination of endogenous processes of technological innovation.[27] Economists Paul Romer[28] and Robert Lucas[29] extended this thought but their focus remained macroeconomic and removed from any specific institutional context. Rosenberg,[30] in contrast, used institutional and historical analysis to investigate the process within the 'black box' of innovation, concluding that 'innovation, the central feature of capitalist reality, is not a product of a decision-making process that can be described or analyzed as "rational" '.[31] He credits Schumpeter with previously asserting that:

> the assumption that business behavior is ideally rational and prompt . . . works tolerably well only within the precincts of tried experience and familiar motive. It breaks down as soon as we leave those precincts and allow the business community under study to be faced by . . . new possibilities of business action which are as yet untried and about which the most complete command of routine teaches nothing.[32]

This historical, institutional approach led a group of scholars to examine the role of path dependency and evolution.[33] The late Keith Pavitt, former director of the Science Policy Research Unit (SPRU) at Sussex University, for instance, argued that much of innovation is uni-directional, based on technology which is 'specific, complex, often tacit, and cumulative in its development'.[34] Political economist Robert Gilpin extends the analysis, noting that the path-dependent nature of most of the major technologies of the day – influenced by the non-rival, excludable nature of the technologies, interlocking networks of learning and knowledge storage and imperfect supply chains – creates the potential for regions or countries to capture first-mover benefits and move to higher growth trajectories.[35] Rosenberg identified a range of path dependent technologies, including steel production, solid state physics and its relationships to the advent of the transistor in 1948, lasers and optical fibers, information theory, biotechnology and material science, all which draw on a range of complementary, co-developing science and technology domains.[36]

A number of alternative models have been offered to explain how innovation actually happens. Etzkowitz and Leydesdorff characterize modern innovation systems as a triple helix of industry, government and university engaged in a collective effort to generate new science and technology.[37] At least three main forms of the Triple Helix model have been identified. In Triple Helix I the three spheres are defined institutionally (university, industry, and government) and interactions across otherwise defended boundaries are mediated by formal industrial liaison, technology transfer, and contract offices. In the second version, the components are defined not as institutions but rather as communication systems, which have the potential to produce new forms of interaction. In Triple Helix III, the university, industry and government perform both their traditional functions as well as assume roles often assumed to be the function of another helix: universities, for example, begin to operate as regional or local innovation organizers.

Professor Michael Gibbons from SPRU and a group of colleagues offered a complementary model, starting with the types of knowledge generated rather than with the institutions developing them.[38] They posit that two modes of knowledge generation now exist: Mode 1, which they call traditional knowledge, is generated within disciplinary, primarily cognitive, contexts; and Mode 2 knowledge is created in broader, 'transdisciplinary' social and economic contexts (Table 3.1). They argue that Mode 2 knowledge tends to lead to a more socially distributed knowledge production system and, hence, is both socially accountable and reflexive. This, they argue, makes Mode 2 knowledge a profound challenge to the traditional governing system because communication tends increasingly to take place across institutional boundaries and not only within established hierarchies. There is some evidence that suggests they may be on to something. A recent analysis of a database of Nobel Prize winners (1873–1998) and great inventors (1900–91) showed that the average age at first invention (which was taken as a proxy for educational attainment) is rising 0.6 years per decade and that the size of the winning US research teams is increasing 17 per cent per decade.[39] This is at least partly due to a rise in specialization, which has risen by 6 per cent per decade over the past century. But this opening up of the research system is not happening at the same pace in all sectors. Recent evidence suggests that drug and medicine inventions (measured through the patent system) cite university science 2.4 times more often than chemical inventions; mechanical patents, in contrast, seldom cite science (only 1 per cent as likely as chemical inventions).[40]

*Table 3.1 Modes of knowledge generation*

| | Mode 1 (traditional) | Mode 2 (new) |
|---|---|---|
| Problems set and generated in | Disciplinary communities | Transdisciplinary economic and social context |
| Nature of communities | Homogeneous | Diverse and heterogeneous |
| Organizational structure | Stable hierarchies | Transient heterarchies |
| Type of quality control | Peer review judgments about the contributions made by individuals | Temporary, heterogeneous sets of practitioners, collaborating in special and localized contexts |

*Source*: Adapted by author from Gibbons *et al.* (1994)

The Committee for Study of Invention disagreed somewhat with Gibbons, in that they see the inventive, knowledge-generation process as fundamentally driven by individuals not institutions. These individuals:

> are not trapped by what they know or think they know. They are boundary transgressors. They mobilize their knowledge flexibly, selectively, and critically. They often abandon what is 'known' in several senses – setting aside previously effective approaches that do not seem helpful in a specific case, bracketing knowledge as not helpful, and challenging prior knowledge as perhaps false or flawed.[41]

Another group of scholars has attempted to define innovation in the context of the steps from the germ of an idea to delivery of a useful good or service. The traditional model was a linear process where basic research led successively to applied research, development, commercialization and use.[42] The simplest of these models assumed the genesis of the initiative was serendipitous – the more sophisticated versions had a feedback loop from users to the basic research stage, where consumer and social demands influenced the areas of research. This model actually worked quite well to describe the type of industrial research labs pioneered by Thomas Edison and his generation of inventors.

Increasingly, though, there is a concern that the innovation process is much more diffuse (analogous to the Mode 2 knowledge envisaged by Gibbons). If one looks at systemic innovation processes (such as the second phase of canola development outlined above), one can see that many inputs and outputs are generated in these newer innovation processes. No single firm or region can truly be viewed as self-sufficient or self-sustaining. Economists Stephen Kline and Nathan Rosenberg, for example, provide a non-linear approach that explicitly identifies the role of both market and research knowledge and the potential for open research systems.[43] Their 'chain-link model of innovation' begins with a basically linear process moving from potential market to invention, design, adaptation and adoption but adds feedback loops from each stage to previous stages and the potential for the innovator to seek out existing knowledge or to undertake or commission research to solve problems in the innovation process. This dynamic model accentuates the role of transacting to find, transfer and use new types of knowledge and information.

The innovation process is complicated because of the interconnections between different phases of the same technological track and the layering of different technology tracks. Our formal data, for better or worse, aggregates across all of these domains, with the result that effective linkages are lost. Further complexity evolves from the fact that firms are frequently involved in projects related to different waves of discovery, such as gene sequencing, genomics or pharmacogenomics (each of which calls for a differentiated set of actors with a specific division of innovative labor). Moreover, within a single wave of discovery, firms frequently engage in several projects that spread over different phases of research.[44] Put into the context of the more conventional lazy-S curve of the life-cycle of a product, firms can be engaged at one or more points on any single curve, while at the same time are involved in multiple different curves with different ranges and amplitudes. This calls for a mix of actors and collaborative agreements.

A number of scholars have examined innovation systems from an array of approaches. Some are interested in the role of innovation systems in minimizing transactional costs[45] or generating economies of scale and scope;[46] others have looked at these systems as an extension of social network theory. A wide range of approaches have been examined, including: industrial/managerial investigations of geographic concentrations of competing and co-operating companies, suppliers, service providers, and associated institutions that create industrial interdependencies;[47] analyses of networks and their relationships between markets and the visible hand of organizational authorities;[48] industrial milieu as a complex exhibiting synergies of economic and technological interdependencies;[49] value chain industrial clusters involving an extended input–output or buyer–supplier chain;[50] comparative institutional analysis;[51] and innovation systems as learning regions acting as collectors and repositories of knowledge and ideas that facilitate the flow of knowledge, ideas and learning.[52] Ultimately, all of these lines of research examine the role of 'untraded interdependencies' – a term now extensively used to refer to an array of externalities of scale and scope.[53] These interdependencies are often characterized as involving region-specific conventions, informal rules and habits that coordinate economic actors in localized production systems.[54] The resulting analyses examine how central actors, networks, culture, alliances and a myriad of other relationships manage uncertainties, thereby lowering transaction costs or increasing the rate and scope of innovative activity. There has been an array of research on specific innovation systems around the world, many of them focusing on biotechnology, an array of ICTs, new media, wine, food, automotive parts, clothing and advanced materials agglomerations in Europe,[55] the US[56] and Canada.[57] Although these studies have shown an array of potentially key relationships between and among firms and other institutions, there is little that one might view as definitively causal or universal.

In spite of extensive research effort, innovation to a great extent still remains a mysterious black-box system. Although we have some understanding of how some features of technology and some social relationships can influence the scale, scope and timing of innovation, virtually all of the models are partial equilibrium models. Each of the models so far uses the *ceteris paribus* condition, assuming for the sake of simplicity that potentially critical relationships will not change in response to innovative pressures. For example, most of the economic analyses assume that the political and social institutions underlying an economy remain constant and social analyses posit that the economic relationships and pathways remain stable. This assumption is clearly not tenable as a starting point, especially when we are examining transformative innovation.

## A GENERAL MODEL OF INNOVATION

A more general model of innovation is needed in order to examine the governing challenges and opportunities. Such a model can be based on a recasting of the work of others. Three points jump out from the above analysis. First, we can definitively say that innovation is clearly different from invention. A prototype fax machine is an invention; the introduction of fax machines into the market is an innovation; and the millionth fax machine in use marks a transformative innovation. Second, there is general agreement that innovation most frequently occurs within cultures, communities and organizations whose aim is to transform information into knowledge, to invent new applications of that knowledge and to convert those creations into socially valued goods, services, processes or organizations. Third, success in many of the partial equilibrium models is marked by the ease in which creations are introduced, absorbed into and persist in the economy and society. Furthermore, we can extrapolate from much of the case study work on specific technologies and innovation systems to conclude that society usually reshapes in a major way what it uses. Seldom is the technology adapted and adopted in exactly the form in which it was conceived. Rather, the advent of a new transformative approach (be it technologically or socially derived) tends to be modified and directed in a wide range of ways that are seldom anticipated and at times are inimical to the interests of the original developer. Hence, change is not a linear pathway, but a cascading set of actions and reactions. The challenge is to understand the norms, rules and processes (either formal or informal) that govern those actions and reactions.

Figure 3.1 offers one approach to a general model of innovation.[58] At its most inclusive, it allows for all assumptions about the scope and governing of innovation to be identified and tested. There are eight identifiable blocks, four stocks of knowledge input/output (the blocks) and four social processes (the ovals). The obvious place to start in the context of the contemporary innovation challenge is with the profusion of information. Information is explicitly

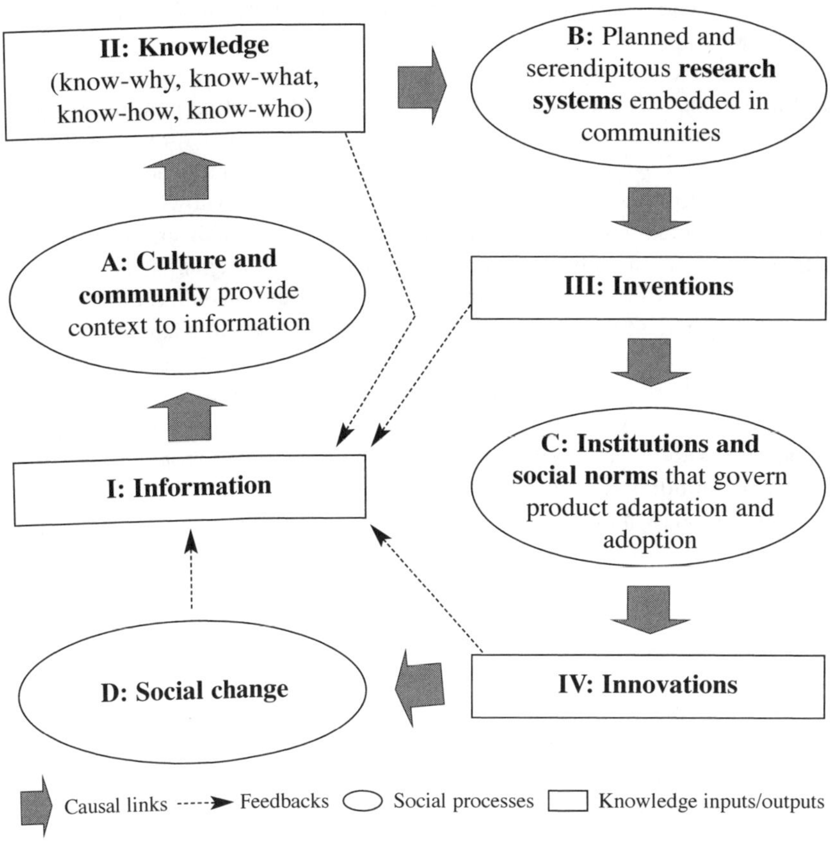


*Figure 3.1 A general model of transformative innovation*

separated from knowledge, in that the latter requires context and increasingly is governed by the complex knowledge 'industry'. At its root, new information, or newly mobilized information, is often the starting point of an innovation process. Formal and informal structures, incentives and processes identify, assemble, transmit and store an array of bits of information in multiple forms. Some information is readily accessible (it is codified and easily searchable) whereas other information may be tacit and only accessible through membership in some closed communities.

Knowledge is generated through the context of cultural and community social processes. Knowledge is fundamentally a social good. To be judged knowledge, more than one person needs to accept it. Although much of the work on innovation has focused on patentable, know-what type knowledge, there is actually a wide range of other types of tacit and codified knowledge that are critical to the generation of an innovation (that is know-why, know-how, know-who and know-when types of knowledge, which are discussed further in Chapter 5).[59]

The creative activity of generating an invention (something that is novel, useful and non-obvious) takes place within a social context and has organizational and social consequences. These three aspects of innovation tend to concentrate innovations in business, organizations and the economy in regional, sectoral or functional clusters in which new knowledge and skills complement imaginative industry leadership, all of which are supported by active partners, including communities and governments. This pattern is frequently seen in innovation corridors, such as found in Silicon Valley in the lower San Francisco Bay area, Boston Route 128, Austin's T3, the Research Triangle Park in North Carolina, Cambridge, Bangalore, and the BioValley spanning the borders of Germany, France and Switzerland.

Translating an invention into a socially embedded innovation then involves a complex web of principals, agents, promoters and regulators on the supply side and middlemen, marketers and consumers on the demand side. Constructing new markets for new products or services is seldom straightforward or simple. Multiple governing systems come to the fore at this stage. The impact of a new technology depends both on improvements in the primitive form and also on complementary inventions.[60] Major technological innovations often necessitate new systems, which are unfathomable at the start. The ultimate impact of new technology is determined by satisfying human needs in ways and contexts that have not yet been articulated.

Finally, as transformative innovations become socially embedded, they begin both to change society and to be changed by society. The end result is that innovations generate new information about our needs, wants, desires, opportunities and constraints. Thus, innovations tend to trigger successive waves of information that can sustain the innovation cycle. Smaller, iterative changes will certainly be changed by society but would be unlikely to have any measurable effect on the structure or operation of society itself.

As mentioned, this model is deliberately highly generalized in order to capture the full array of potential points where activities may need to be governed. Clearly, if one is talking about discrete, one-off technological introductions, this model could be reduced by ignoring some of the feedback and taking as given (*ceteris paribus*) some of the background or context for the governing systems. This model then could collapse to the linear, chain link, or lifecycle models used by others. In that sense, this framework is not intended to replace those other configurations, but rather to amplify our understandings of the underlying institutions, norms, conventions and processes that either nurture or impede innovation.

## NOTES

1. Amabile *et al.* (1996: 1154–5).
2. See Phillips and Khachatourians (2001) for a detailed analysis of the canola story.
3. NRC (1982), McLeod (1982).
4. See Phillips and Khachatourians (2001) for details.
5. Schumpeter (1934).
6. Schumpeter (1939: 84).
7. Rosenberg (1994: 4–5).
8. Rogers (2003).
9. Ruttan (2005).
10. For example, Woodman *et al.* (1993); Amabile *et al.* (1996).
11. Metcalfe (1995).
12. Committee for Study of Invention (2004).
13. For example, Romer (1990); Grossman and Helpman (1991); Krugman (1998).
14. OECD (1997).
15. For example, Metcalfe (1995).
16. For example, Malbera and Orsenigo (2005).
17. For example, Freeman (1987); Lundvall (1992).
18. For example, Stern *et al.* (2002).
19. Etzkowitz and Leydesdorff (1995).
20. Hicks (1932).
21. Rosenberg (1997).
22. Rosenberg (1982).
23. Rosenberg (1994).
24. See Alston *et al.* (1995) for an overview of the theory and practice.
25. See Alston *et al.* (1998) for a review of studies in the agri-food sector.
26. Alston *et al.* (1995).
27. Arrow (1962).
28. Romer (1990).
29. Lucas (1998).
30. Rosenberg (1982, 1994).
31. Rosenberg (1994: 53).
32. Rosenberg (1994), citing Schumpeter (1939: 98–9).
33. For example, Dosi and Nelson (1994); Nelson (1995); and Saviotti (1996).
34. Cited in Acha *et al.* (2004).
35. Gilpin (2001).

36. Rosenberg (1994).
37. Etzkowitz and Leydesdorff (1995).
38. Gibbons *et al.* (1994).
39. Jones (2004).
40. Hall (2004).
41. Committee for Study of Invention (2004: 16).
42. See Rogers (1983).
43. Kline and Rosenberg (1986).
44. Nesta and Mangematin (2004).
45. For example, Phillips (2005).
46. For example, Porter (1990).
47. For example, Porter (1990).
48. For example, Powell (1990).
49. For example, Maillat (1991: 113).
50. For example, Bergman and Feser (2001).
51. For example, Peter Maskell and Lorenzen (2003).
52. For example, Florida (2002).
53. Introduced by Dosi (1984) and Lundvall (1988, 1992).
54. Storper (1997: 5).
55. For example, Asheim and Gertler (2005), Cooke (2002); Porter (1990).
56. For example, Saxian (1994); Porter (1990); Audretsch and Feldman (1996).
57. Wolfe and Gertler (2004); Phillips (2001).
58. Leydesdorff (2005) offers a somewhat different but analogous treatment of the interrelationship between knowledge, innovation, geography and the economy.
59. Mokyr (2002a) breaks down knowledge into propositional knowledge (also called 'know-why') and prospective knowledge (which appears to be a combination of 'know-what' and 'know-how').
60. Rosenberg (1994).

# 4. Governing

**Govern, v. 1**. To rule with authority, esp. with the authority of a sovereign; to direct and control the actions and affairs of (a people, a state or its members), whether despotically or constitutionally; to rule or regulate the affairs of (a body of men, corporation) . . . To direct and control (a person, the members of a household) with the authority of a superior . . .

**OED**

## INTRODUCTION

Many observers of the information and telecommunications revolution either boast about or bemoan the active governing of the new technology. Depending on the observer, the rapid emergence and dominance of new technologies has been a marvel of the unrestricted market or the result of errors of omission and commission by governments at all levels and in all markets. Putting rhetoric aside, it is reasonable to inquire about how these apparent global transformations come about.

At one level, the emergence of waves of new information technologies and applications has appeared to be the result of unfettered markets. Individual entrepreneurs have identified market needs (or anticipated where demand might be generated) and cobbled together packages of equipment, software and content to push into the market. Many of the most prominent advances were extremely informal, involving self-financed, out-of-work engineers tinkering in their garages (such as David Hewlett and William Packard, who established Hewlett-Packard Company in Packard's garage with an initial capital investment of $538), teenagers working in their parents' basements (Linux was initially written as a hobby by Finnish university student Linus Torvalds while attending the University of Helsinki), venturing entrepreneurs (Trip Hawkins left his job at Apple Computer in 1982 and used a personal investment of US$200 000 to found what was to become Electronic Arts, the world's largest video games publisher) or the venture capital market (many of the dot-coms in the 1990s emerged, were hyped by venture capitalists, ballooned and vanished without producing any revenues). In each case, there appeared to be little or no overt government involvement. Furthermore, the founders and promoters of these ICT ventures created myths that their founding, and continuing existence, is a testament to pioneering, individual enterprise. They assert that, not only did they not have any government support, they were seldom regulated or constrained in their operations. In short, they are the epitome of entrepreneurship and the free market.

The reality falls far short of the myth. Formal oversight and investment by local, regional, national and international public, governmental organizations has been instrumental in creating demand and managing the overall industry while less formal but still vital private and collective structures have created the conditions for individual initiative to flourish. Perhaps most important of all, governments everywhere have until very recently either directly owned or formally controlled through legislation and regulation the vital telephone infrastructure, access to that infrastructure and the content that can use those systems. Almost all of the national and regional telephone systems outside North America were state-owned utilities, often acting as proponents or champions for their local industrial partners or as strategic buyers of tailored technologies. In North America, a hybrid system evolved, with some local or regional systems directly owned by the state but with national regulatory oversight of the rest (sometimes regulating price and access but always regulating the potential for competition). Increasingly national utilities are being privatized, but they remain under the formal oversight of telecommunications regulators. Meanwhile, national governments have variously directly owned, controlled or regulated the content on the airwaves (radio and television), using them to set and enforce standards for taste, obscenity, culture and national content. Although the state has had less of a say in the evolution of the content of the Internet, governments have still been instrumental in making the system work. Robert Kahn, one of the two individuals credited with the genesis of the Internet, offers a detailed review of the role of the US government, noting that the US Department of Defense through the Advanced Research Projects Agency initiated the work that led to the TCP/IP protocols that allow data packet switching between different kinds of machines and the DOD adopted the TCP/IP as its standard protocol, that the ARPA established and chaired the Internet Configuration Control Board and its successor committees to govern design of the system, and that the US Department of Commerce issued a sole-sourced contract in 1998 to the Internet Corporation for Assigned Names and Numbers (ICANN), a California non-profit corporation, to manage the assignment of domain names and IP addresses.[1] Without these institutions, the Internet would not have emerged as quickly or widely as it has.

Recently there has been significant international debate about how the Internet should be governed. On 30 June 2005 the US announced, unexpectedly, that it will retain authority over ICANN, which handles the assignment of names and the underlying internet protocol numbers.[2] Its regulatory mandate flows from two agreements between ICANN and the Department of Commerce: an annual MOU[3] which designates ICANN de facto control of the legacy A-root, preeminent server in Hendon, Virginia, and a contract for ICANN to administer the Internet Assigned Numbers Authority (IANA). There is some concern that the Authority lacks the capacity to properly manage the system. A recent assessment of disputes through the IANA which involved issues of free speech and commercial domain suggests that these disputes tend to be resolved by panelists without any formal adjudication training and, as a result, the panels tend to side with commercial interests, without the appropriate use of hearings, the opportunity to examine parties or to call for further evidence.[4]

Numerous other market and non-market institutions have been involved in establishing the technical standards, protocols and procedures that govern the introduction of new technologies in this area. This management has even extended to the social acceptability of the system – for example, many transportation companies, gasoline vendors, stores, restaurants, movie chains and libraries have individually or collectively imposed rules of approved usage of mobile phones. At times the state has also intervened – an increasing number of jurisdictions are legislating to restrict the usage of all but hands-free cellular telephones while driving.

This oversight extends far beyond what the literature suggests. Calestous Juma, a former secretary general of the Convention on Biological Diversity, Harvard professor and sometime advisor to UN Secretary General Kofi Annan, recently reported on a delightful fusion of new technologies and traditional governing models. He had been traveling to his home in rural Kenya, at Port Victoria on the shores of Lake Victoria, where he discovered that mobile phone access generally remained quite spotty. In absence of any managed solution, local papers report that individuals have found that they can resolve their access problems by strapping a mobile phone on top of a stepladder and then, whenever they hear the phone ringing, climbing up the ladder (fast of course) to receive their calls. He discovered, however, that people increasingly were finding new approaches to use their existing infrastructure. He reported that local people generally knew where to pick up a strong signal – and would quickly disseminate knowledge of 'high bast' locations. One source was up trees which were traditionally used as sites for communicating with others. One such tree was near Juma's grandmother's house. He reports: 'I had to wait in line, of course – such traffic should be expected. The place is carefully managed – balancing between privacy and the capacity of the branches to support the users . . . In most cases privacy is not an issue because much of the use here is text messaging.' He also discovered that other solutions were being sought.

> In one popular spot they have planted a designated tree (socially referenced because it is described in relation to the owners of the neighboring property – sometime quite confusing because of the use of terms as 'where so-and-so used to live' or 'where-the-Great-Tree-used-to-be'), which is obviously baffling for those whose local history is outdated. Their GIS system is clearly very sophisticated, alive and robust. One of the mobile phone providers (Kencell) recently installed a booster tower in the area. Ironically, my grandmother's tree is next to the tower, which is owned by a competing firm, which is my provider (Safaricom). Right under their nose . . . Africa seems to be getting onto the information highway, step-by-step. And every time a tree is cut down, someone is cut off.[5]

The evolution of the information and communications revolutions and the efforts of a wide range of actors seeking to govern infrastructure, content, access and use warrants more detailed examination. This brief exploration of the ICT revolution suggests there are multiple ways for existing and new actors to use traditional and purpose built systems to govern the development and use of a new technology. It may, in fact, be this fusing of old with new that signals the emergence of a successful transformative technology.

## WHAT DOES GOVERNING INVOLVE?

The question of how complex changes are governed is extensively discussed in the literature. Three different groups start from differing perspectives – political scientists define the problem as one of regulating the economy and society; economists focus on the role of institutions; and sociologists and post-modernist scholars talk about governance by a diverse set of actors. Each of the different domains offers a different take on how we govern our economy and society. Political scientists tend to view it as a question of how government can command, control and order relationships in markets or society. Economists see the world comprised of an array of transactions that are fundamentally grounded on self-interested optimization. Sociologists tend to see the world as composed of social groups that are seeking to frame the world in terms of their values. These roots are examined further in Chapters 5–8, but for the time being it is important to keep in mind that each perspective has a qualitatively different view of the governance challenge.

One way to get a reading on the scale of the challenge of governing transformative technologies is to review the literature on regulation, institutions and governance.

### Regulation

The political science literature is grounded on the search for order. Kimon Valaskakis, the former Canadian Ambassador to the OECD, defines government 'as the exercise of authority or the distribution of power. In essence it is related to the notions of order and decision making. The opposite of governance is anarchy, or the absence of order.'[6] Thinkers and writers have long examined how government can establish and enforce rules in pursuit of public goals. Governments can be and often are defined as regimes or systems of laws, administrative rules, judicial rulings and practices that constrain, prescribe, and enable government activity.[7]

Governing through rules is fundamentally about power – 'the ability to do or effect something or anything, or to act upon a person or thing'.[8] The difficulty of dealing with the concept of power, however, is that there are two competing views of power. First, there is the 'ability' or capacity to compel action or conformity and second there is the concept of power as a measure of 'political or national strength'. The blending of those two concepts in much of the literature and discussion since the creation of the modern nation state (via the Treaty of Westphalia of 1648) has tended to cause political scientists to focus on the role of the state in exercising and conserving power, rather than looking at all actors and the power they hold or exercise.

In recent years there has been an effort within the political science field to expand the concept of regulation. The literature has looked at the evolving role of private industry in corporatist systems, at the role of communities or networks and at different configurations of power and authority.

Domestic and international politics both offer illustrations of an evolving role for private firms in public affairs. There is a wide body of literature and debate about the corporatist or neocorporatist concept of formally engaging unelected corporations representing economic, industrial and professional groups in the legislation and regulation of the state. This approach, drawing on roots laid down in the medieval guilds, was developed and propounded in nineteenth- and twentieth-century Italy and Germany as an alternative either to a market based system or to socialism. Generally corporatist states recognize dependence on corporate actors but also seek to use them as an instrument to both effect and legitimate policies. More recently, the literature has identified and decried a creeping corporatism in liberal democracies, through lobbying, regulatory capture and the sometimes dominant role of the military industrial complex. Neocorporatism has emerged more recently as a set of new social arrangements involving tri-partite bargaining between unions, the private sector, and government. Various attempts have been made to bring the three parties together nationally (in many European countries, such as post war Germany) or internationally to deal with wage bargaining and to facilitate major economic transitions. One area that many point to as an area of emerging corporatism is in the management of international trade and the international financial system; recent analyses suggest that rather than a true corporatist model emerging, we may actually be moving toward a multilayered system of governing in these areas.[9]

Canadian political scientists Michael Atkinson and William Coleman suggest that policy networks (generally described as corporatist, state-directed, collaborative or pluralist) and policy communities are perhaps better ways to look at some realms of regulation, because they explicitly target many interwoven and overlapping actors and relationships beyond the political-bureaucratic system.[10] Community specifically refers to the actors whereas the network refers to the relationships among actors. They posit that networks involve relationships of power and dependency that 'transcend and color individual transactions' and thus require more than transactional economic analysis. The implication is that the dichotomy between the state and any other actor will not adequately frame any analysis. Rather, the appropriate framing should be a collection of actors in a more broadly conceived policy arena. Regime theory has evolved and built upon this network/community approach, focusing on explaining how events are regulated by governments in concert with others.

The scope of study of command and control structures can be further widened to address the broader concept of 'governance' from sociology. The challenge is to examine ruling systems at all levels of human activity – from the family to international organization. In that context, governance encompasses both the activities of governments, and the many other channels through which commands flow in the form of 'goals framed, directives issued, and policies pursued'.[11] Thus, global regulation involves geographic, social, cultural, economic and political processes that transcend boundaries. Once disaggregated, global governing then, as political scientist James Rosenau asserts, is 'the sum of myriad – literally millions of – control mechanisms driven by different histories, goals, structures, and processes'.[12] A major distinction between Rosenau and the sociological approach is that he still is focused on command and control events, only he is now willing to consider actors other than government exercising the power; in contrast, the sociology-based governance literature goes beyond formal command and control situations, and considers the exercise of norms, convention and cultures (which lack any sense of command and control features) that orient actors to conform rather than direct them to comply. He cautions, however, that as we broaden the analysis to include new actors with new motivations, there may not be any characteristics or attributes common to any of the resulting mechanisms.

University of Quebec political scientist Gilles Paquet suggests we face new forms of distributed governing arrangements 'based on a more diffused pattern of power and a new valence for various meso-systems'.[13] Distributed governing 'does not simply mean a process of dispersion of power toward localized decision making within each sector: it entails a dispersion of power over a wide variety of actors and groups'.[14] A seemingly infinite array of distributions of power evolves in response to pressures to adjust to rapid change. 'The fabric of these new "worlds" is defined by the new dominant logic of subsidiarity in all dimensions: it welds together assets, skills and capabilities into complex temporary communities that are as much territories of the mind as anything that can be represented by a grid map.'[15]

Economist Herbert Simon examined the concept of complexity in the context of regulation and similarly concluded that:

> [T]he interest in recent years of many sciences in complexity and complex systems has drawn attention to the fact that most of the complex systems seen in the world are nearly decomposable systems. They are arranged in levels, the elements at each lower level being subdivisions of the elements at the level above. Molecules are composed of atoms, atoms of electrons and nuclei, electrons and nuclei of elementary particles. Multicelled organisms are composed of organs, organs of tissues, tissues of cells . . . Near decomposability is a means of securing the benefits of coordination while holding down its costs by an appropriate division of labor among subunits . . . nearly decomposable systems will adapt to the changing environment and gain in fitness more rapidly than systems without this property.[16]

The political science literature offers a number of important concepts for understanding the challenge of governing. The sustained focus on command and control systems as the defining feature of governing systems is important, in that it offers a clear point of delineation from both the economics and sociology literatures. Second, as the literature has become much less state-centric, it has introduced the potentially useful concepts of multilayered, disaggregated, distributed or decomposable governing systems, which are the opposite of the homogenizing, oligopolistic outcomes envisaged in the corporatist model.

## Institutions

Economists envisage the world as made up of economic beings that seek to optimize their welfare through production and consumption. A modern economy and society is then characterized as an agglomeration of transactions. Provided the underlying assumptions of the economic model hold – scarcity, full information, free entry and exit, no market power and no externalities of production or consumption – the unaided, voluntary transaction between two ungoverned, but self-centered individuals could yield an optimal volume and distribution of goods and services. In a 'perfect' world, then, all relationships or exchanges that conform to the strict economic assumptions could be governed by relatively liberal markets. Some economists[17] have gone so far as to examine key social institutions, such as marriage, through the economic frame, explicitly ignoring the role for either overt state rules or socially defined considerations. Much of economics today thus remains isolated from any debate about governing.

British philosopher Karl Polanyi was one of the early modern scholars to reassert that the market is not an organic, autonomous creation but is rather 'embedded' in and subordinated to politics, religion and social relations. He noted that before the end of feudalism, all economic systems were organized 'on the principle of reciprocity or redistribution, or householding, or some combination of the three'.[18] The artificial separation of the market from society can be traced to the works of Adam Smith, Jeremy Bentham and Alfred Marshall. But he argues that the tendency to 'truck, barter and trade' is not enough to explain the emergence of markets. Polanyi argues that the nature of markets is the result of 'highly artificial stimulants'. He asserts 'there was nothing natural about laissez-faire; free markets could never have come into being merely by allowing things to take their course'.[19]

The innovation analysts and growth theorists are the single most significant exception to that orientation. Rosenberg notes that Schumpeter, often credited as the father of innovation studies, had a fundamentally different view of the nature of the economy. Whereas the bulk of modern economics is about equilibrium analysis, or at most attempting to look inside the Walrasian process of establishing equilibrium, Schumpeter argues in *Capitalism, Socialism and Democracy* that 'the essential point to grasp is that in dealing with capitalism we are dealing with an evolution process . . . Capitalism . . . is by nature a form or method of economic change and not only never is but never can be stationary.'[20] In short, the central tendency of capitalist economic systems is disequilibrium, which poses profound challenges for our concept of the role of the state, and others, in governing the economy and innovation.

Innovation, because it creates conditions of disequilibrium, requires an array of institutions to emerge and flourish.[21] Five state-defined or at least state-supported institutions are thought to have been most important in the innovation processes in the United States – limited liability, marketability of common shares, stock markets, insurance and defined obligations between principals and agents. A critical function of these systems is to provide innovators with some certainty about government involvement by constraining governments from acting arbitrarily.[22] Which specific institutions are needed is far from clear, but the need for some is well accepted.

Nobel Prize winning economist Douglass North provides the classical economic definition of institutions. They are:

> the humanly devised constraints that structure political, economic and social interaction. They consist of both informal constraints (sanctions, taboos, customs, traditions, and codes of conduct) and formal rules (constitutions, laws, property rights). Throughout history, institutions have been devised by human beings to create order and reduce uncertainty in exchange. Together with the standard constraints of economics they define the choice set and therefore determine transaction and production costs and hence the profitability and feasibility of engaging in economic

> activity. They evolve incrementally, connecting the past with the present and the future; history in consequence is largely a story of institutional evolution in which the historical performance of economies can only be understood as a part of a sequential story. Institutions provide the incentive structure of an economy; as that structure evolves, it shapes the direction of economic change towards growth, stagnation, or decline.[23]

This mirrors the all-encompassing definitions of governing and regulation used by political scientists and, as will be shown, mirrors the prevailing view of governance espoused by sociologists. Although many economists adopt and cite this definition, they seldom make much use of its inclusive nature.

Vernon Ruttan offers an alternative, narrower and more economically based definition of institution – he suggests institutions should be defined to exclude the informal codes and norms that comprise the cultural endowment of a society and instead focus on the rules or conventions that govern behavior within, among and between economic units and their environment. This definition focuses in the same way as the emerging political definition on the role of the institution on effecting specific behaviors. Cultures, norms and informal codes, though perhaps foundational, cannot be traced directly to control of specific behaviors; or if they can, they are often encased in and governed by some form of incentive or enforcement regime that is explicitly designed to guide outcomes.

## Governance

Sociologists in recent years have dominated the debate about how we govern our economy and society. They have introduced the concept of governance to encompass all the varied methods of governing, ranging from state regulation, market institutions and a wide range of civil authorities. There is an array of concepts embodied in the term 'governance'. There is some evidence that the concept of governance has been used for many centuries to refer to the exercise of authority in generally apolitical organizations, such as firms and universities. The concept generally implies the existence of a political process that involves building consensus or obtaining consent or acquiescence of groups of citizens with diverging interests, in order to carry out a proactive policy or program.[24] Government and governance traditionally were used as synonyms.[25] The roots of the differentiated concept of governance were eclectic, coming from development studies, political science, and organization studies.

One can trace the underlying context of a multilayered concept of global governance to the 1970s and the rising dissatisfaction with the state-dominated model of development.[26] The first reported use of the term 'governance' formally surfaced in its current and modern form in 1989 in a World Bank report.[27] There would appear to have been an effort by development agencies to create a new 'rhetoric' of international, interpersonal, social and political relations, involving a broader set of actors and variables. The array of definitions offered by the development community – the World Bank, UNDP, OECD, Commission on Governance, UN Secretary General, International Institute of Administrative Sciences and the Tokyo Institute of Technology – suggests that governance was purposefully defined (that is focused on one or more of economic and social development, good government, respect for human rights and public life) as the use (such as a manner or process) of power (both political and private use of both relational and structural forms) and authority (both political and social) through a variety of distributed mechanisms, both formal (processes, laws, institutions) and tacit (norms, values), that ultimately mediate differences (over interests and values).[28]

Political scientists see the concept as still inextricably tied either to the institutions of government, or at least the functions of governing. For example, although some accept that although governance is sometimes used in the sense of steering and control activities in different spheres of society (such as for profit enterprises and voluntary organizations), it 'refers principally to the exercise of authority in government and the political arena'.[29] Similarly, although governance involves the 'general exercise' of authority which often goes beyond government and the political arena, the state remains a critical actor.[30] Nevertheless, the important change is that the determination of the objectives, laws and methods of governing has largely moved away from 'absolutism, authoritarianism and even the autarkic conception of the modern state'.[31] This has come about because of: struggles to introduce greater democracy and competitive markets; greater productivity and wealth in the hands of individuals; changes in the management of firms, governments and households; and the expansion of intergovernmental organizations.

Nevertheless, there remains some ambiguity in the definitions. The concept seems to be evolving as an almost government-neutral concept. One recent definition posits that 'governance generally refers to the means for achieving direction, control and coordination of wholly or partially autonomous individuals or organizations on behalf of interests to which they jointly contribute'.[32] Although government is implicit in this conception, it is not dominant or even central to it.

British political scientist Rod Rhodes suggests that governance should not be viewed as simply a recasting of analysis, but actually 'a new process of governing; or a changed order condition of ordered rule; or the new method by which society is governed'.[33] As such, it is not a synonym for government but rather involves a new system of 'self-organizing networks or "governing without government" '. He identified six functional uses of the term governance in the period between 1989 and 1995, showing the fecundity of the field. Governance could entail: a minimal state with a new corporate government, such as openness; new public management, involving professional management (as advocated and adopted in many bureaucracies); a shift from rowing to steering;[34] the advocacy of 'good' governance (used by the World Bank and many national development agencies); a socio-cybernetic system which is distributed and 'centerless' (involving subsidiarity, absence of a single sovereign authority, multiplicity of actors, interdependence and blurred boundaries); and self-organizing networks, involving interdependent actors that cannot be steered.

Governance in such a world would then refer to self-organizing, inter-organizational networks which exhibit interdependent, sustained game-like exchanges where the interactions are rooted in trust. In essence, central authority is replaced by self-organizing networks. Complex systems have been defined as involving a large number of elements, building blocks or agents, capable of interacting in a wide variety of ways with each other and with their environment.[35] Interactions may occur with either immediate neighbors or distant ones and agents can be identical or different and may move in space or occupy fixed positions. The common characteristic of all complex systems is that they display organization without any obvious external organizing principle being applied. Even simple systems may produce complex, often non-linear, outputs, which is the epitome of 'chaos'. Such systems, though hard to predict, exhibit high degrees of adaptability. Network theory, in particular, offers insights into the various distributed authorities in a world of subsidiarity. Networks define the permutations of actors (or nodes) through relationships (connections or links), where networks of different density can have either central nodes or distributed relationships (the nature and stability, or volatility, of networks can often be modeled through small, discrete behavioral models, such as the 'game of life' cellular automata model). Analysis of a variety of small world networks, such as power grids, ecosystems and epidemics has revealed that most human-built systems are seldom the result of a single design but rather are part of an evolutionary process of competing, merging and mutating self-organized sub-routines.

It is important to note that these definitions are not simply nested views of the governing problem, but in many fundamental ways differ in their assumptions and conclusions. For example, the distributed, centerless model is not the same as networks; the centerless model is based on subsidiarity, which implies a top-down decision to devolve or vacate a policy or program area in favor of another actor, while the network model implies the initiative comes bottom up without any higher coordination.

Sociologist Bob Jessop explores the roles of markets, states and partnerships, focusing on the heptarchic governance of economic, political and social life. He sees governance as 'any mode of co-ordination of interdependent activities'.[36] He explicitly identifies three modes: 'the anarchy of exchange, organizational hierarchy and self organizing "heterarchy" '. He observes that inter-systemic coordination is typically de-centered and pluralistic, with heterarchic or self-organizing governance based on reflexive rationality, which relies on continued dialogue and exchange of information. Others have called this dialogue the function of 'voice'.[37]

Increasingly there is agreement that governance does not rest on the authority and sanctions of government.[38] The concept refers instead to a set of institutions and actors drawn from government and beyond, involves blurring of boundaries and responsibilities, entails power dependence between institutions and encompasses autonomous self-governing networks of actors and works without the power and authority of government. The Commission on Global Governance went on to define governance as:

> the sum of the many ways individuals and institutions, public and private, manage their common affairs. It is a continuing process through which conflicting or diverse interests may be accommodated and co-operative action may be taken. It includes formal institutions and regimes empowered to enforce compliance, as well as informal arrangements that people and institutions either have agreed or perceive to be in their interest.[39]

The concept of governance thus does not represent a single, consistent view of how we govern. Rather, it offers an array of ways of looking at governing. Many of the concepts, such as the notions of centerless and networked governance, offer conflicting views. Nevertheless, the combination of the various views offers some insights into the modern system of governing. In the first instance, the concept clearly suggests that there are alternatives to governing through the authority of the state or relying on the market – the voluntary or civic sector offers an alternative, heptarchic way of governing. Second, these alternative systems are increasingly interdependent, which has the potential to challenge traditional notions of sovereignty and accountability. Third, the diffusion of power and authority has the potential to blur the boundaries between the state and market (this book examines in later chapters ways to increase precision about the models of governing in the three domains – public, private and collective – to reveal the increasingly finely drawn boundaries between the domains).

## CONCLUSION

Ultimately, governing involves a number of critical elements: a purpose; a set of actors (the governors and the governed); a domain to be governed; and a process of governing. A place to start, then, is with the concept of governing. The OED defines the verb 'governing' as involving swaying, ruling, influencing, guiding, directing or regulating the course of an event. This can be in human affairs or in nature. Governors thus may be mechanical devices (such as the governors on early steam engines), laws of nature (including the role of genes in determining the heredity of traits) or any human-made institution (such as the criminal code and common law) that influences an outcome and the governed may be nature, machines, people, firms and societies. The processes of governing thus involve a wide array of domains, actors and processes. It is becoming increasingly clear that one cannot simply think of governing as what states or governments do; rather, governing involves the array of humanly constructed processes that determine how different interactions will lead to specific outcomes. The concept of governing is far bigger than the laws, regulations and actions engaged in by state governments.

Although governing involves all human constructs – both those that are formally and consciously constructed and adhered to and those that are informal and often part of our background consciousness – this analysis focuses on those active forms of governing where there is a direct correlation between the governing system and the outcomes. The focus then is on those processes that involve controlling, directing or regulating human affairs. Many would contest that approach, arguing instead that governing should also include the many intangibles that underpin society – the conventions and customs acquired over the millennia. Though these factors are important, their import goes way beyond their role in governing systems. Hence, these factors are for the most part excluded from the analysis.

The challenge of governing is not a new problem. During periods of slow, iterative change the underlying, embedded, governing structures of modern economies are not easily discerned. But during periods of major disruptive change, their role and function becomes clearer. Two types of change have revealed the importance of governing systems: liberalization in command and control economies; and periods of technological change.[40]

Technological revolutions, in particular, have challenged our governing structures. Nathan Rosenberg argues that the first industrial revolution (1760–1830) was only possible because the institutional negative feedback of the earlier period was removed. He argues that:

> the freedom to conduct experiments is essential to any society that has a serious commitment to technological innovation or to improved productive capacity. The starting point is that there are many things that cannot be known in advance or deduced from some set of first principles. Only the opportunity to try out alternatives, with respect to both technology and to form and size of organization, can produce socially useful answers to a bewildering array of questions that are continually occurring in industrial (and industrializing) societies.[41]

The advent of knowledge-based economies has raised new concerns about who's in charge of governing the economy. In the knowledge economy, the key asset is the ability to innovate – the facility to develop, adopt and adapt new ideas, products, and organizational structures by combining existing ideas, products and structures in new ways. Ultimately, this process involves the identification, assembly and use of disparate types of information and knowledge through a wide range of social governing systems. Political scientist Lester Salamon suggests that finding the right tools to effectively govern in this environment will be difficult because these knowledge networks are plurifom (diverse), self-referential, asymmetrically interdependent and dynamic, and consequently do not share the same goals, operating styles, skills, worldviews, incentives and priorities.[42]

Although Rosenau expresses reservations about the ability to create typologies and to identify commonalities among subgroups of governing systems, the literature offers some places to start. The three convergent approaches – regulation, institutions and governance – provide useful insights into the nature of the challenge and offer a few hints on how we can bring greater clarity to analysis. In the first instance, there would appear to be three domains of governing (state, markets and civil society), which operate on somewhat different aspects of life (political, economic and social) and use different governing mechanisms (command and control, exchange and voluntary association). That much would appear to be agreed by many. Though it is clear that the resulting governing systems are distributed, multilayered and interdependent, it is less clear whether they are fundamentally self-organizing networks or are centerless governing systems defined by subsidiarity and mutual recognition. Resolving this issue will go a long way to sharpening the oft-cited fuzzy boundaries of state, market and civil society.

## NOTES

1. Kahn (1994).
2. Economist, The (2005) 'Eminent domain', 9 July, p. 56.
3. http://www.icann.org/general.
4. Murray (2003).
5. Personal correspondence with Juma, 2004.
6. Valaskakis (2001: 46).
7. Tarschys (2001: 3).
8. OED On-line.
9. Held *et al.* (1999).
10. Atkinson and Coleman (1992).
11. Rosenau (1995: 15).
12. Rosenau (1995: 16).
13. Paquet (2001), citing Elkins (1995) and Storper (1996).
14. Paquet (2001: 190).
15. Ibid.

16. Simon (2000: 753).
17. For example, Nobel Prize winner Gary Becker.
18. Polanyi (1944: 57).
19. Polanyi (1944: 145).
20. Schumpeter (1954: 82).
21. Rosenberg (1994).
22. Rosenberg (1994: 98).
23. North (1991: 97).
24. De Alcantara (1998).
25. Stoker (1998).
26. Weiss (2000).
27. Pagden (1998).
28. Weiss (2000).
29. Tarschys (2001: 28).
30. Michalski *et al.* (2001: 9).
31. Ibid., 9.
32. Lynn *et al.* (2000: 2).
33. Rhodes (1995: 1).
34. As observed and advocated by Osborne and Gaebler (1992).
35. Amaral and Ottino (2004: 148).
36. Jessop (1998).
37. For example, Picciotti (1995) and Kaufmann *et al.* (2004).
38. Stoker (1998: 17).
39. Commission on Global Governance (1995: 2).
40. Stiglitz (2002) argues that the rough and often damaging transition from authoritarian to liberal market economies in many former Soviet controlled areas was the result of a failure of the 'Washington Consensus' to appreciate the political and social context of the transition economies. Western advisors simply failed to understand the basis for the exchange economy. Tarschys (2001) concurs, asserting that the economic collapse of the Soviet Union and the differential economic performance in Africa (elaborated in the World Bank's 1997 World Development Report) are the result of a lack of understanding of the underlying governing systems. Handelman (1994: 89) offers a fascinating illustration of this, noting that more than 40 per cent of the turnover in goods and services in Russia by 1993 was controlled by organized crime because 'Russian policymakers committed a fundamental mistake: they tried to develop a free market before constructing a civil society in which such a market could operate'. Stiglitz (2002) sees more problems on the horizon as many of the problems of globalization can be attributed to the governance models currently being used: command and control state systems have been discredited; the commercial/economic paradigm alienates many; and there are limited alternatives for governing collectively.
41. Rosenberg (1994: 88).
42. Salamon (2002a).

# PART III

# Tools for analysis

# 5. A framework for analysis

And so these men of Indostan disputed loud and long,
Each in his own opinion exceeding stiff and strong.
Though each was partly in the right, they all were in the wrong!

**John Godfrey Saxe**[1]

## INTRODUCTION

The parable of the Blind Men and the Elephant, attributed to the Pali Buddhist Udana and compiled in the second century BC, is as good a place as any to start a discussion of a framework for governing transformative innovation. The parable recounts that a community of blind men once heard of an extraordinary beast called an elephant and resolved to understand the creature. They sent a number of wise men from their group as emissaries to find the elephant and return with information. Once they found the animal, they each took a part of the animal to feel. One touched its leg, the other a tusk, the third an ear and they left believing they knew the animal. When they were questioned by their fellow blind men, they each offered different interpretations. The one who had felt the leg believed the animal was a pillar, extremely rough to the touch, and yet strangely soft. The one who had felt the tusk described the elephant as hard and smooth, with nothing soft or rough about it; furthermore, the animal was not nearly as stout as a pillar, but rather had the shape of a post. The third, who had held the ear, described it as both soft and rough, but not like a post or a pillar, but rather like a large piece of leather. Each was right in a certain sense, since each of them described the part of the elephant he had examined, but none was able to describe the entire animal.[2]

This parable has been used over the millennia to describe and illuminate our limited ability to develop descriptions and proofs of our human condition. In a way, it remains an excellent way to discuss the methods of investigating human governing systems. If we take the current debate about governing transformative technologies, we have an analogous condition to the elephant parable.

The 'blind man' who begins with a government perspective is concerned about creating, managing and maintaining governing systems that are accountable, responsible, and transparent (what LSE scholar Julie Black labeled

'ART').[3] There are many good analyses of the structures and challenges of states governing new technologies.[4] This literature tends to focus on the direct and delegated uses of formal power. The emerging consensus in many quarters is that these systems are stressed and unable to handle many of the varying demands and pressures but that non-governmental alternatives fail the ART test.

Meanwhile, 'blind men' with an economic or market focus concentrate on the transactions that facilitate production and distribution, and tend to see all critical factors being represented in either the transaction itself (where the marginal cost equals the marginal utility of the marginal unit consumed) or the whole host of factors that underpin and facilitate arm's-length, impersonal exchange. There is a large and growing body of literature about how consumer opinions are formed,[5] of how firms make choices about how to structure their markets[6] and about how value is generated in the process of exchange.[7] This literature tends to conclude that the unassisted market can and does perform most, but not all, of the essential governing functions.

Finally, 'blind men' with a sociological bent or who propound non-governmental collective management of new technologies tend to concentrate upon how or to what degree processes engage civil society or decisions reflect the divergent views in society. There are a range of scholars who have adopted this approach.[8] Their focus is on how or whether the governing system has the right process or the right outcomes, and tends to be critical of the lack of 'reflexivity' in both the state and market governing systems.

The challenge is to find a paradigm that links to each of the realities examined by the respective 'blind men'. Though Rosenau expressed reservations about accomplishing this, we have the advantage that in the context of governing transformative innovations we are dealing with explicit human constructs that are rooted in public, private or collective systems in society.

## GROUNDING INTERDISCIPLINARY STUDY

Interdisciplinary studies, such as into governing transformative innovation, are inherently messy. In the first instance, interdisciplinary analyses generally are not derived from theory but rather are triggered by observations of society or nature. Thus, the very nature of defining an interdisciplinary problem tends to be a normative exercise. Moreover, the structure and focus of any interdisciplinary analysis tends to be grounded in the definition of the problem itself, rather than in any particularly coherent methodology. This leads to a problem of finding a clear scope for the analysis and a single, generally acceptable methodology for undertaking the work. Many critics assert this leads to a lack of method.

The OED defines method simply as a 'special form of procedure, esp. in any branch of mental activity'; methodology thus is 'orderly arrangements of ideas; orderliness, regular habits'. Method or methodology will thus vary widely among different disciplines and approaches, as they each use a different array of rules, postulates and data. Because the method chosen is often inextricably entwined with a discipline and a researcher's approach, it is often a hidden or overlooked aspect of many studies. The choices the researchers and analysts make about a research topic – why they want to examine a topic, what aspects they will examine, how they will examine it and what data and evidence they will use to advance their arguments – often predetermines the results and affects their ability to accurately reflect likely events or to influence debate.

This brings us back to the blind wise men. Ultimately, what researchers and policy analysts examine and what they conclude is heavily influenced by the underlying motivations of the researcher. Some investigators and disciplines are inherently normative, in that they are driven by an ethical or normative judgment of 'what ought to be'. These studies often start with a perspective of wanting to change the world, and then identify the area of analysis and define the methods in ways that further that goal. As such, normative studies frequently are challenged as being biased and incomplete. Much of the work done on transformative innovation seems to at least partly reflect a normative bias. Many of those who have investigated the topic appear to have a driving concern that something is wrong with the distribution of power and authority in our current governing system. Though not a fatal flaw, it tends to divert analysts' attentions from a full, clear and complete examination of the current operating system and instead to focus their attention and effort more on identifying and examining options for change.

In contrast, many disciplines and investigators strive to remove the ethical and normative perspective from their work, instead seeking to pursue an objective, 'positivist' examination of 'what is'. Most of the theoretically based social and natural science disciplines have gone toward the positivist approach, seeking to find the interconnections between elements and to examine dispassionately how new or changed circumstances might influence other factors. For example, the investigations by molecular biologists and plant scientists into the genetic code and tools to modify the code of life can involve significant ethical and moral considerations, but the scientific investigations can and usually are undertaken without reference to those aspects. Instead, researchers in these areas are simply trying to understand cause and effect.

The distinction between positive and normative studies is perhaps more apparent than real. Some philosophers argue that at root almost all analysis has a normative component.[9] In the first instance, researchers are usually drawn to understand a topic because of some personal desire to see their knowledge used. Similarly, an investigation may begin with a normative aim but quite often the favored tools and approaches of positive theories are adapted to the investigation.

Another area of disconnect is in the choice of research approach. Many people distinguish between two basic kinds of logical argument: induction and deduction.[10] Induction is usually described as moving from the specific to the general, or more particularly it is the process of arriving at a conclusion based on a set of observations. It is finding patterns from a complex environment. Inductive arguments often include analyses that assume the whole is like the individual parts, extrapolations to areas beyond the study (which are assumed to be like those studied) and predictions into the future based on the past. Though not viewed as 'logically valid' by mathematicians, many political scientists, sociologists and others tend to follow this approach to examine real world phenomena. Deduction, in contrast, begins with general theory, laws, rules or other widely accepted principles and uses them to deduce a specific event or structure. Deductive reasoning thus refers to the process of concluding that something must be true because it is a special case of a general principle that is known to be true. Deductive reasoning is therefore 'logically valid' and is the fundamental method by which scientific, mathematical and many social science (especially economic) theories and laws are derived and sustained. Deductive theory also tends toward reductionist analysis, as the causes and effects are more finely parsed and tested.

In practice, any inductive argument can be expressed deductively, and vice versa. Regardless of how one initiates a study, it is possible to include both inductive and deductive approaches. Difficulty arises because different disciplines may be intolerant of alternative approaches. For example, many social scientists think economics is far too deductive and reductionist to be a valid starting point for social analysis whereas economists tend to view other social scientists as weak in method and therefore 'softer' in approach and structure. This all-or-nothing dichotomy is counterproductive. There are clearly opportunities for hard-core deductive, reductionist analysis in the study of causal relationships in the economy, government and society but there are also circumstances where the inherent simplicity of 'elegant' deductive models simply fails to capture the nature of a relationship. The goal of this chapter is to tease apart this dichotomy to identify where the different theories, models and approaches can offer the greatest insight into the challenges of governing transformative innovation.

## KNOWLEDGE SYSTEMS

Transformational innovations are fundamentally based on applying knowledge. Given that, it is important to have a better understanding of what knowledge is and how it can be used to transform the economy or society and how it can be governed.

The late Professor Susan Strange, one of the key scholars who defined the nature of international political economy, posited that the place to start such a discussion is with what she called the knowledge structure which 'determines what knowledge is discovered, how it is stored, and who communicates it by what means to whom and on what terms . . . Power and authority are conferred on those occupying key decision-making positions in the knowledge structure.' [11]

Knowledge displays mercury-like properties – we can see it but it is hard to grab on to. The difficulty is that knowledge exhibits an array of dichotomies: it can be both explicit and implicit; it can be local or global; and it can be individual or collective.[12] This array of attributes means the functional space encompassed by the term is wide.

There have been various attempts through the millennia to parse the concept of knowledge into more tractable components. Aristotle, for example, offered three classes of knowledge: universal and theoretical knowledge; instrumental and practice-related knowledge; and normative, common sense-based knowledge. Alternatively, knowledge could be parsed based on its degree of codification and formalization.[13] Codified knowledge (similar to Aristotle's universal and theoretical knowledge) is systematic and reproducible whereas tacit knowledge (those skills and corresponding understandings that defy articulation or codification and correspond to Aristotle's instrumental and normative knowledge types) is considered intangible – residing in individuals or communities. Knowledge could also be categorized based on it primary purpose: 'propositional' knowledge relates to natural regularities and phenomena (also called 'knowledge what'); and 'prescriptive' knowledge consists of instructions that constitute techniques or routines (or 'knowledge how to').[14]

There are at least four discernable types of knowledge: know-why, know-what, know-how and know-who.[15] One might conclude that the four possibilities emerge from combining the dyads of form (codified or tacit) and structure (whether residing in individuals or the collective). Know-why knowledge is codified and collective, know-what knowledge is codified and individual, know-how is tacit and individual and know-who is tacit and collective (Table 5.1).

Each type of knowledge has specific features.[16] 'Know-why' refers to scientific knowledge of the principles and laws of nature, which for the most part is undertaken globally in publicly funded universities and not-for-profit research institutes and is subsequently codified and published in academic or professional journals. This knowledge would be in the knowledge block in the chain-link model, having been created almost exclusively in the research block. 'Know-what' refers to knowledge about facts and techniques, which

*Table 5.1 Classification of types of knowledge*

| | Degree of Codification | Produced by | Extent of disclosure |
|---|---|---|---|
| Know-why | Completely codified | Universities and public labs | Fully disclosed or published in scientific journals |
| Know-what | Completely codified | Universities, public labs and private companies | Privately held but fully disclosed in patents |
| Know-how | Tacit, not codified | Hands-on in labs or teams | Limited dispersion; often due to trade secrets |
| Know-who | Tacit, not codified | Exists within firms or research communities | Limited to community |

*Source*: Adapted by author from Malecki (1997), p. 58.

can usually be codified and transferred through the commercial marketplace. The stock of know-what is in the knowledge block in the chain-link model, having been created in the research, invention, design and adoption blocks. 'Know-how' refers to the combination of intellectual, educational and physical dexterity, skills and analytical capacity to design a hypothesis-driven protocol with a set of expected outcomes, which involves the ability of scientists to effectively combine the know-why and know-what to innovate. This capacity is often learned through education and technical training and perfected by doing, which in part generates a degree of difficulty for the uninitiated and makes it more difficult to transfer to others and, hence, more difficult to codify. Know-how would be represented in the research block and also in the invention, design and adaptation stages. Finally, 'know-who', which involves information about 'who knows what and who knows how to do what'[17] is becoming increasingly important. As the breadth of knowledge required to innovate expands, it has become absolutely necessary to collaborate. In today's context, know-who also requires knowledge of and access to private sector knowledge generators who at times may hold back the flow of crucial or enabling information, expertise and knowledge. Know-who knowledge is seldom codified but accumulates often within an organization or, at times, in communities where there is a cluster of public and private entities that are all engaged in the same type of research and development, often exchange technologies, research materials and resources and pursue common staff training or cross-training opportunities. The arrows in the chain-link model would represent this type of knowledge, as building relationships that lead to trusting networks of know-who is the basis for those flows.

This categorization of knowledge goes part way to addressing Strange's challenge to find and focus on those who occupy key decision-making positions that determine what is discovered, how it is stored, and who communicates it by what means to whom and on what terms. In the first instance, the taxonomy reveals that there are both stocks and flows of knowledge, which would suggest that there will need to be correspondingly different actors responsible for governing their different functions. Furthermore, no one type of individual or institution will control all of the knowledge and its uses. Rather, there will be an array of individuals, public agencies, private enterprises and collective structures that will govern different types of knowledge, each potentially using a different set of processes and seeking a different set of objectives.

## COMPETING GOVERNING SYSTEMS

Fundamentally, we are faced with a complex systems problem. There is no single approach to understanding the dimensions of the challenge of governing transformative technology. Many other economists, sociologists and political scientists have accepted this necessity and have attempted to map out a more interdisciplinary framework. Each discipline has made tentative efforts to consider and accommodate actors not normally considered in their domain.

A number of economists have noted the non-market collective efforts that underpin critical aspects of the market. Nobel economist Herbert Simon, for example, argues that although Adam Smith's invisible hand appears to secure coordination without obvious central planning, 'this invisibility of mutual dependence is deceptive. The usefulness of markets depends on a *shared* knowledge.'[18] He notes that price information is often supplemented with extra-transactional information (such as quality reviews and product standards). As a result, we live in an organization economy, which exhibits a predominance of organizational over market activity.[19] Given this, he posits that maintaining economic equilibrium cannot be left to the invisible hand of the market but rather requires government action.[20] This is particularly true when faced with dynamic phenomena and uncertainty in large, complex social systems.

Growth theorists have been particularly interested in the role of non-rival goods.[21] Sharing of non-rival goods, and possibly even some rivalrous goods, is becoming an increasingly important component of competitive markets. Non-rivalrous goods are relatively easy to share, as there are no congestion costs related to sharing them. Some rivalrous goods, such as excess capacity on home computers, may also lend themselves to networking and sharing.

Institutional economists have also been interested in the role of non-market actors. Albert Hirschman, an institutional and development economist, posited that loyalty, exit and voice form the foundational trilogy of institutional economics.[22] He particularly saw voice as fundamentally a political or collective process whereas exit and loyalty are part of the economic exchange system. No system could survive without all three. Hence, markets, except in very narrow static analysis, cannot be viewed as independent of other governing processes. Others have characterized the problem in the context of the underlying nature of goods or services: whether they are rivalrous or not; whether they can be made excludable; and whether voice is needed to deliver them.[23] Different balances of these factors will lead to different costs for searching, negotiating and enforcing transactions through the marketplace. Particular institutions – the government sector, the private sector, the participatory sector and four hybrids of these 'pure' types – tend to be best suited to govern particular types of transactions.[24] Each domain represents different individuals, involves different incentives and is effective in producing goods or attributes with specific characteristics.

Those political scientists who look at state-to-state relationships in the context of globalization have increasingly been forced to consider new repositories and wielders of power. Strange, for example, sees power as the critical factor that 'determines the relationship between authority and market.'[25] Her view is that all power can be distilled down to one of three roots: coercive, overt force; economic wealth; and moral or social ideas. Different groups derive different sorts of power from different sources. There are two types of power: relational and structural power. Relational power, conventionally described by realists as the power of A to get B to do something they would not otherwise do, is often the focus of international relations. Structural power, in contrast, is the power to shape and determine the structures of the global political economy within which others have to operate.[26] The difference is that structural power is not overtly coercive; rather it is the ability to determine how relationships and transactions shall be managed. The capacity to exert either relational or structural power is a function of three interrelated factors: first, the root of the power being exerted (force, wealth or ideas); second, four power 'structures' (the security apparatus, the financial system, the production system and the knowledge system); and third, the array of 'bargains' that mobilize and normalize actions and reactions in the use of power. In much of her work, Strange focuses on the structural power of nation states, but near the end of her career she raised the possibility that non-state actors, such as the self-governing professions like accountancy and the academies, may increasingly exert structural power that is either a support for or competitor to national capacity.

Meanwhile, those analysts who focus on intra-state relations are similarly finding new actors warranting consideration. New forms of distributed governing arrangements are emerging, based on a more diffused pattern of power (sometimes called 'meso-systems').[27] Distributed governing does not simply mean dispersion of power toward local decision making but rather involves dispersion of power over a wide variety of actors and groups. New power systems facilitate more effective and efficient learning needed during times of rapid change – decentralized and flexible sub-routines based on 'moral contracts' and reciprocal obligations negotiated in the context of evolving partnerships. Thus, 'the fabric of these new "worlds" is defined by the new dominant logic of subsidiarity in all dimensions: it welds together assets, skills and capabilities into complex temporary communities that are as much territories of the mind as anything that can be represented by a grid map.'[28] This heterarchical system is a world without a pecking order. Instead, it displays 'strange loops' of authority, much like the game of paper, rock, scissors, where no one authority dominates in all circumstances, but rather dominates only under limited conditions.[29]

Economist Kenneth Boulding argues that human relationships can be classified as the compulsory, the contractual, and the familistic.[30] The compulsory system involves threats (such as 'you do something for me or I'll do something nasty to you') which fundamentally depend on the credibility of the threat and the capability of the partners to affect their sides of the relationship. These types of arrangement could yield either zero-sum or negative-sum results. The contractual, exchange-based system involves voluntary bids, offers and contracting, which generally yield positive-sum outcomes for both parties. The familistic, integrative system involves status relationships (where 'you do something because I am your father, or a king, a priest, a teacher, a lover, child, student . . .'). This yields three different methods of integration: coercive relations that distribute rights and obligations, led by the state; quid pro quo exchanges in the market governed by the Marshallian scissors of supply and demand; and voluntary dealings, where cooperation, reciprocity and solidarity engage community and society.[31] Boulding argues that society can be viewed as a triangle (his 'social triangle'), where all organizations – including the state, the market and civil authorities – are built on one or a balance of the three relationship systems.[32]

Meanwhile, sociologists have engaged in the analysis of the post-industrial, knowledge-based world through a variety of multi-dimensional framing devices, such as the triple-helix framing device.[33] Generally, they believe that relational power is important in a post-modern liberal democracy but that it is only effective if it conforms to the underlying norms of the society. In other words, non-state actors have structural power that can offset or reduce relational power. 'Mode-2' type knowledge, in particular, appears to be created, owned, controlled and used through an array of formal and informal relationships among three main actors – governments, industry and universities.[34] Each has specific roles, interests and capacities to facilitate knowledge-based growth. More recently, analysts and theorists have attempted to explain the rise of non-governmental, civil or social action groups, some at the local level and some at the international level. The Mode 2 framework could be modified to incorporate a role for other civil authorities in the context of one of the three main actors or by adding a new fourth strand in the helix to represent society.[35]

The array of different approaches can be distilled down to three functional types: governments that structure rights and obligations; markets that engage in voluntary economic exchange; and a variety of collegial or social groups – including universities and NGOs – that engage in social discourse and mobilize voice in society.

## INVESTIGATING COMPETING GOVERNING SYSTEMS

The institutional challenge of governing transformative innovation can thus be conceived of as the result of the interrelationships between three differently conceived domains. First, the nature of the activity being governed provides the foundation for the analysis. Second, the goals, objectives and central directive for the governing agency will define the manner of the relationships that evolve. Third, governing happens in some space, which suggests some geographic context needs to be considered.

First, the transactional approach to governing offers some insights into a useful way of categorizing economic and many social activities. That approach suggests most traded goods or services exhibit three main characteristics – rivalry, excludability and voice – which can be used to identify three categories of pure goods – public, private and collective.[36] Rivalry is simply a measure of how many people can use a property at the same time. If a property can only be used by one person at a time (such as a hammer), then the product has a high level of rivalry; if it can be used simultaneously by more then one user at a time (such as a computer program), it is non-rival. The more disembodied the property, the less rivalrous it is likely to be. National defense and public health are often cited as examples of purely non-rivalrous goods and services. Excludability is a measure of whether people can be kept from using the good or service. The degree of excludability depends on the physical structure of the property (land may be fenced but ideas cannot), the cost of duplicating and using the property and the legal and social impediments to use.

Highly excludable products, like the clothes on your back or the bathroom in your home often occupy space whereas those outputs that are hard to exclude people from (once again national defense and public health) tend to have wide application and universal access. Voice measures the amount of input required from others to generate value from a good or service. The more production or consumption depends on information provided by others, the more it is said the product has high voice.

Thus, going out to buy a physical good like a hammer does not really require much discussion and common agreement, whereas defining what an organic food standard entails would require significant discussion among buyers, sellers and producers to ensure the standard has meaning. If we can get beyond the economic slant of the terminology, we can see that these three concepts can offer insights into many public and social activities. These three factors can be used to categorize three relatively discrete types of pure goods and a multiplicity of hybrid goods with mixed properties. Public goods (such as national defense and public health) can be defined as those activities that are characterized by low excludability, low rivalry and limited voice. Although some democratic debate may be involved in the decision to produce a public good, they are inherently designed for the entire population and not simply those mobilized in any consultation. Private goods (such as machinery and equipment and most consumer products) usually exhibit high excludability, high rivalry and limited voice. Common pool goods (such as conservation, standards setting and marketing services) involve high voice and may involve a range of degrees of excludability or rivalry.

Second, a comprehensive triad of institutions – governments or states, the market and social or familial organizations – form the basis for all governing systems.[37] Governments have multiplied aggressively in recent years. After millennia of limited government or government by empire, we saw an explosion in the number and orders of governments in the past century. A tabulation of the data in the *CIA Factbook* shows that as of October 2005 there were 192 independent states and 73 dependencies, areas of special sovereignty, and other miscellaneous entities and that these states rely upon another 4010 first-order administrative divisions inside their territories.[38] Although there are no definitive numbers of other orders of government, a ball park estimate is that there may be as many as 4 or 5 million local governments, including 437 mega cities each with more than 1 million inhabitants.[39] Meanwhile, at the international level, these 192 states have developed and participate in an estimated 6415 international governmental organizations.[40] Each of these governing entities is structured to assign rights and to enforce obligations within their areas of jurisdiction. They primarily exert relational power through an array of coercive powers, ranging from military and civilian force, to legal and regulatory processes, to taxation. Beyond the provision of basic rights and obligations (which one might argue exhibit non-rivalry and non-excludability), governments are most focused on providing public goods that exhibit non-rivalry, low excludability and limited need for voice.

Markets are also pervasive. By some measures, everyone in the world is engaged in multiple ways in the US$55.5 trillion global market, either as a consumer, a producer, an employee, an owner, a lender or a borrower. That means that there are at least 6 billion potential actors in this domain. For the sake of efficiency and effectiveness, we have structured our producing and financial intermediation into a relatively small number of enterprises. By some counts there are around 1.5 billion economic units producing some goods or services, including 1.34 billion individuals economically active in farming (one of the more basic economic activities in many parts of the world)[41] and up to 300 million commercial firms operating around the world.[42] Although the characteristics of these different economic actors undoubtedly vary widely, they exhibit some common attributes. Each of them is engaging voluntarily (or at least with some degrees of willingness) in the act of economic exchange (involving goods, services, land, labor or capital). At root, voluntary exchanges involve a number of common activities, such as search, negotiation and enforcement of contractual arrangements. Exclusion is a fundamental feature of markets – individuals who lack the resources (labor, capital, land or purchasing power) are by constitutional design excluded. Economic theory suggests that self-satisfying individuals, in pursuit of their own self-interest, if unobstructed (and subject to their own constraints of time, assets and preferences) will optimize their production, consumption, savings and investments, such that the market economy will deliver the maximum utility and profit to the society for any given resource endowments. These self-directing actions have been shown to be very powerful in markets where all of the benefits and costs of a decision to produce or consume are embodied in the transaction. Markets will be increasingly less effective as the degree of externalities that cannot be accommodated in the terms of the transaction rises.

Meanwhile, there is increasing attention to a growing number of non-governmental organizations and various social groups that are forming and engaging in social debate. The LSE Group on Global Civil Society estimates that there were more than 59 000 international NGOs in 2003, each attempting to mobilize the voice of a group in society that shares a common value or interest. This domain includes a wide range of groups, including social action groups with concerns related to economic development, infrastructure, research, the environment, social development, law, policy and advocacy. These groups are fundamentally defined by their ability and desire to reflect the views of their members or supporters, which leads them to exhibit a wide range of structures and governing systems. They may be highly exclusionary and provide relatively rivalrous services (or vice versa) but they founder as collective agencies if they lose touch with the values and interests of their members.

Each specific institutional structure tends to be more effective at producing particular types of goods.[43] The government sector is best at producing public goods – the low excludability makes privatization infeasible and the low voice component makes it difficult for the collective sector to organize. The private sector tends to dominate whenever property rights can be assigned to make rival goods excludable. The property of exclusion allows private firms to sell at the marginal cost of production. The participatory sector is best at governing common pool goods (such as marketing services) – the collective group will usually have more information that will enable them to more effectively manage the resource and capture the benefits.

Governing innovation thus ultimately involves both economic and non-economic actors and incentives. Who leads will depend on the nature of the output to be governed: rivalry (does the governance model create congestion costs?); excludability (can people be kept from benefiting from the governance system?); and voice (do governors require specialized, intermediate inputs from others?). Depending on the dominance or balance of these factors, public state, private market or collective authorities may offer the most effective loci for governance.

This balancing act, however, does not happen in a vacuum. Rather, these institutions compete at the local/regional, national and international level. There are: discrete local, regional, national and international state-based structures; local and global liberal economic markets; and an array of communal associations, clubs or social action groups that span the range from local to global.

## CONCLUSION

We have seen that a number of social processes are producing increasingly greater complexity – competition and interdependence in the economy and society creates new roles for actors to fill (sometimes called co-evolutionary diversity); actors are at times being encouraged to add new subsystems (structural deepening); and large systems are made up of smaller simpler subroutines (capturing software). These complex systems inevitably involve variably coupled (from loosely to tightly), interdependent, multiple entities within a dense web of causal connections that is open to external events. The synergies within these open systems may both exhibit non-linear behaviors and generate synergies and feedback that may stimulate learning. Canadian political scientist Thomas Homer-Dixon cautions us to avoid getting carried away with the sense of apparent chaos. He notes that although all human systems can exhibit some chaotic behavior, 'unlike natural systems, humans can learn, so they can change their behavior to avoid chaos. Although it is true that many social systems are extraordinarily complex and non-linear, they are not wholly disorganized systems; rather, they exhibit at least partial self-organization.'[44] The result is frequently an ever-changing pattern of new actors, playing new roles, often with less transparency of power, authority, accountability and responsibility.

The next three chapters examine the theoretical and analytical foundations of the study of states and government (Chapter 6), the market (Chapter 7) and social and civil authorities (Chapter 8).

## NOTES

1. Saxe, J. (n.d.) The Parable of the Blind Men and the Elephant. Retrieved from the world-wide web on 3 October 2005, at: http://www.kheper.net/topics/blind_men_and_elephant/Sufi.html.
2. Fritz Meier *The Problem of Nature in the Esoteric Monism of Islam* from *Eranos Yearbooks* cited at http://www.kheper.net/topics/blind_men_and_ elephant/Sufi.html.
3. Black (2002).
4. For example, Doern and Reed (2000), Isaac (2002); Vogel and Kagan (2004).
5. See Evenson and Santaniello (2004).
6. For example, Shapiro and Varian (1999); Kalaitzanondakes (2003); Rausser and Goodhue (2003); Scotchmer (2004).
7. For example, Alston *et al.* (1995).
8. For example, Beck (1992); Bauer and Gaskell (2002).
9. For example, Brunk *et al.* (1991).
10. Arthur (1994)
11. Strange (1992: 121).
12. Jensen *et al.* (2004).
13. Polanyi (1962).
14. Mokyr (2002a).
15. Lundvall and Johnson (1994); OECD (1996); Malecki (1997). Boulding (1984, 1968), talked of three types of knowledge – know-what, which creates know-how artifacts and know-whether, which involves evaluation and decision making.
16. See OECD (1996) for a good overview.
17. OECD (1996).
18. Simon (2000: 750) (emphasis added).
19. Ibid., 751.
20. Ibid., 755.
21. For example, Romer (1990).
22. Hirschman (1970).
23. For example, Mahoney (1992); Picciotti (1995); Gray *et al.* (2006).
24. Mahoney (1992); Picciotto (1995).
25. Strange (1994; 23).
26. Ibid., 24–5.
27. Paquet (2001).
28. Ibid., 190.
29. Paquet (2001), citing Ogilvy (1986–7).
30. Boulding (1970), citing Polanyi, 1957, and Sorokin, 1959.
31. Paquet (2001).

32. Boulding (1970: 30).
33. Ettzkowitz and Leydersdorff (1995).
34. Gibbons *et al.* (1994).
35. Mehta (2003).
36. Picciotto (1995).
37. Boulding (1970); Paquet (2001) and Benacek (2005) each use a form of the Boulding triangle, with the corners dominated by the three dominant institutions.
38. CIA (2005) notes that first order administrative divisions is a concept developed and approved by the US Board on Geographic Names and includes states, provinces, territories, districts, departments, counties, parishes, boroughs, governorates, communes, self-governing cities, municipalities, cantons, or emirates.
39. http://www.citypopulation.de/World.html
40. http://www.uia.org/statistics/organizations/ytb199.php
41. FAO FAOSTAT estimates for 2003.
42. Calculated by applying the North American ratio of non-farm firms operating relative to the total population (about 4.7 per cent) to the global population.
43. Picciotto (1995), drawing on Coase (1937) and Boulding (1970).
44. Homer-Dixon (2001: 125).

# 6. Governing via the state

> No matter how much we believe in our institutions and in the regularized procedures of our societies, no matter how just, rational, and durable we think them, they are at best only loosely grounded on some form of bedrock reality or immutable truths that endure beyond human beings. To a considerable degree, they are sustained by collective belief and consensus, by tacit, unquestioned, and often grossly simplistic assumptions about how the world works, and often by mutual and willful self-delusion. Our societies cohere and function in no small part because most of us want them to cohere and function, and because the alternatives are, for must of us, literally unthinkable.
>
> **Thomas Homer-Dixon**[1]

## INTRODUCTION

While some are concerned about the apparent retreat of the state from governing in some areas, others argue that we have too much governing by the state. When it comes to grappling with the role for the state, one can find justifications for both views. Since the dissolution of the empires, we have seen a large increase in the number and orders of government operating around the world. With more than 192 independent states, 4010 first-order administrative divisions, an estimated 4 or 5 million local governments and 6415 international governmental organizations, we certainly are not short of examples of state-driven governing. Each state system represents a discrete slice of society that can and usually does differ by geography, history, race, social stratification, cultural attributes, economic interests or functional activities. Different governments, faced with similar situations, thus may choose to pursue radically different policies, strategies and programs.

Examining only a few examples shows the diversity of state government. In spite of talk about the homogenization of society and culture through globalization, there is perhaps the widest range of government systems ever recorded. While an increasing number of peoples have chosen a post-enlightenment, liberal, democratic, pluralist system of representative government, we still have some indigenous communities in the Americas and Africa that sustain hunter-gatherer societies through traditional kinship systems that trace back to pre-recorded times. In the spectrum between those two extremes, we have examples of almost all forms of government tried in the past three millennia, including absolute and constitutional monarchies, confederations, federal and unitary states, participatory democracy, dictatorships, communist or socialist regimes, theocracies, autocracies, empires, warlords and tribal systems. We even have territories that lack any recognizable state government, and from the outside those regions would appear to border on anarchy, where life, as philosopher Thomas Hobbes asserted, really is 'solitary, poor, nasty, brutish and short'.[2]

These states, as one might expect, engage in an almost complete spectrum of activities. A few activities are common to all systems. At root, states (or other institutions acting in the role of states) are the only institutions that have shown any sustained ability to exercise overt, coercive power. In most cases, states use this power to secure the territorial integrity of their country and to maintain order inside the country, generally through the enactment and enforcement of criminal and civil laws. The level of coercion exercised varies widely. Some states engage in a determining way in every facet of the economy, society and personal life; others allow individuals and institutions the maximum amount of freedom, only engaging when domestic order is seriously threatened.

Beyond arranging for security and order, the array of actual tasks undertaken by states is almost limitless. We can find examples where states either currently, or in the not-so-distant past, have directly or indirectly engaged in ordering almost every aspect of the economy, society and even family life. People living in long-time liberal, democratic countries often forget that virtually every single economic activity that we rely on markets to provide could be and currently is somewhere managed and coordinated through non-market systems. The means of production (especially land and capital), for example, are usually collectivized or monopolized by authoritarian socialist states and most indigenous communities; similarly, many developing and even some advanced industrial economies have at times extensively managed specific parts of capital and land markets, defining who can own, control or use available resources, almost always using non-price mechanisms to secure supply and to allocate that supply to potential users. Labor markets have similarly been organized along non-economic lines. While most countries have gone beyond the institution of slavery, indentured servitude still survives in many countries. Similarly administrative edict is still used in many jurisdictions to apportion educational opportunities, to allocate jobs or to define how individuals may use their time in economic affairs. Although many advanced industrial countries have moved away from direct management of the input markets, they remain critical actors, creating order in the markets by regulating and promoting the structure and outcomes using a variety of incentives and penalties.

Moving from the input markets to look at output markets, we can see the long arm of the state in many areas. Central planners in a wide range of government systems (ranging from theocracies, to warlords, to some advanced industrial democracies) often engage in limited or global management of the economy, allocating inputs, directing production and distributing outputs through some non-market way, completely ignoring utility maximization or profit optimization. Farming, fishing and trapping, for example, are frequently collectivized or centrally managed; as we can see in the recent rounds of the WTO, even in areas where farms are privately managed, farm policies and markets are highly regulated, managed and distorted with production, consumption and export subsidies and taxes. Forestry and mining production in many countries remains state owned, controlled and managed, sometimes using market-based approaches and sometimes using administrative edicts. In spite of significant privatization, a majority of states or their lower orders of government probably still own and operate as public utilities large parts of their critical infrastructure, including transportation (such as national airlines, national and regional airports, ports, coastal and deepwater fleets, railways, highways, city streets, buses and local transit systems), communications (including telephones, satellite systems, television and radio broadcasters, broadband wire and wireless transmission systems) and miscellaneous utilities (such as water distribution and purification systems, irrigation systems and power generation and distribution). Many manufacturing and processing sectors have also been under government control at some point in many countries. Similarly, in more centrally administered countries, wholesale, retail and distribution systems, banking, insurance and real estate have either been precluded from operating on market-based rules or have been directly owned, controlled or managed. In brief, virtually every type of commercial activity can and has been delivered directly by the state somewhere in recent history.

Beyond the market, governments have been active in defining, shaping and controlling society. In the first instance, the state has become the single largest generator of economic activity in most advanced economies on the basis of promoting, providing, funding and regulating the delivery of health, education and social services. Based on that area alone, most states command and direct more than one-third of their nation's economic resources, often with limited or no reliance on the market-based price system. Furthermore, governments have rules that define the rights and obligations of citizenship and the level of autonomy an individual may have to marry, have a family, engage in economic affairs or undertake social and democratic activities (variably based on mental competence, age, gender, racial background, educational attainment, wealth and social status). These measures may be coercive rules, such as under apartheid or autocratic systems, or they may be embedded in a set of incentives and barriers that has the same effect of delimiting an individual's liberty.

Even in advanced, industrial democracies, we have fairly intrusive rules about sexuality (prohibiting or criminalizing homosexuality and extramarital sex), reproduction (many countries ban or carefully proscribe abortion, contraception, surrogates, family size, sex selection, and any form of human embryonic stem cell science), marriage and other social unions (ranging from permitting to prohibiting interracial, intra-family, same-sex or polygamous civil or religious unions), intra-family relationships (provisions and regulatory processes relating to how children, the infirm and incompetent may be treated) and even death (including defining the legal and medical conditions for death, the dignity and disposition of dead bodies and the dissolution and distribution of a deceased person's estate).

When it comes to governing transformative innovation, different states have approached the challenge differently. It is probably fair to say that a variety of approaches that span the spectrum have been tried. In some cases the state has used its coercive power to prohibit any development or use of a new technology. In others, it has been the owner, developer and sole proprietor of the technology. In even others, it has engaged in a delicate balancing act of encouraging innovation but overseeing the risks and liabilities that might evolve from the innovation. Given that transformative technologies span the spectrum of human existence, are incorporated in many economic sectors, influence significant parts of our social and personal life and generate significant political debate, it is difficult to isolate those aspects of governing that are unique to the state from those that can and perhaps should be undertaken in other domains.

If we are to understand what the appropriate role of the state might be in governing transformative innovation, we are going to be forced to go beyond the inductive approach of looking at examples of government and then generalizing from them. Because some government somewhere is likely involved in almost every aspect of human existence – spanning all economic, social, cultural and personal aspects – there are simply too many permutations and combinations to gain much insight via that route. Rather, we will need to understand the strengths and diversity of governing via the state by adopting a philosophic, deductive approach, looking for assumptions about the nature of man and the source of authority to govern.

## GOVERNING BY ASSUMPTIONS

There is no generally accepted view of where to start a discussion about government. As noted in Chapter 5, one might start by distinguishing between states, markets and authorities. If markets involve voluntary, mutually beneficial exchanges, governed by self-interest, and civil authorities are defined by kinship relationships directed to forming and promoting a set of values, then, at its root, one might conceive of the state as engaged in coercive relations that distribute rights and obligations. One could characterize the state as a system of hierarchies driven by constitutional or political democracy and focused on coercive administration (what some call 'non-equivalent, enforced transfers') to effect distributions of wealth and rents.[3]

Each governing system is thus fundamentally based on certain assumptions – the market on self-interested individuals engaging in an optimization exercise using economic calculus; civil authorities engaging in a social process seeking to preserve or promote their shared values, interests or epistemology; and the state in its pursuit of order. Whereas the assumptions underlying the role of both the market and civil authorities can be assessed through a deductive process of reduction and experimentation, the assumptions underlying the state are not testable. Economist Vernon Ruttan concludes that 'what is absent . . . is a theory of fundamental forces . . . that brought us to our political condition and continues to push us through political turns'.[4] Hence, the alternate assumptions about the state are inevitably going to be open to heated debates about their validity and appropriateness.

At the lowest common denominator, state systems are defined and, to some extent, determined by their perceptions on the nature of man. Philosophers have spent a great deal of time attempting to delineate the different conceptions of man and their implications on government. Although certainly a simplistic view, one might argue that there are two distinct philosophically justified approaches to governing by the state (there are obviously some state governments that decline to justify their existence other than by might). One philosophical thread involves a pre-Reformation view of mankind still found in theocracies and in the vestiges of many constitutional monarchies. This approach assumes that man is a fallible creation of God (a deity or another higher being) and that all power and authority resides with that higher authority. The higher authority then chooses, through often unclear or little justified processes, temporal and spiritual rulers to govern people.[5] As such, power flows from the top downwards to people, but ultimately rulers in such systems are fundamentally accountable and responsible to God, and not those they govern. In such a worldview, humans are subservient to an all-powerful, non-responsible higher authority. This view underpins most early forms of government, virtually all pre-Enlightenment states, many continuing theocracies and many autocratic governments.

In contrast, the liberal tradition of the Enlightenment introduced two new theories. First, the 'natural rights theory' asserts that human beings are not chattels but have inalienable, natural rights to life; and a 'theory of social contract' that asserts that government is the product of humans transferring some of their rights to a collective. Thus, even though God made man, all power is vested in man. The inalienable right to self-determination is the foundation for representative or delegated government. Individuals collectively delegate authority to some higher authority, but that authority remains responsible and accountable to the people.[6]

Government structures will then depend on which approach is taken. One way to get a sense of the assumptions underpinning a state is to read the preambles to the national constitutions. Article 2 of the constitution of Iran, an example of a theocracy, asserts that:

> The Islamic Republic is a system based on belief in: (1) the One God (as stated in the phrase 'There is no god except Allah'), His exclusive sovereignty and the right to legislate, and the necessity of submission to His commands; (2) Divine revelation and its fundamental role in setting forth the laws.

The US constitution, in contrast, lays out the foundation for humanist, republican government. The US Declaration of Independence of 1776, the foundational document of the Republic, begins with the declaration that:

> We hold these truths to be self-evident, that all men are created equal, that they are endowed by their Creator with certain unalienable Rights, that among these are Life, Liberty and the pursuit of Happiness. That to secure these rights, Governments are instituted among Men, deriving their just Powers from the consent of the governed, – That whenever any Form of Government becomes destructive of these ends, it is the Right of the People to alter or to abolish it, and to institute new Government, laying its foundation on such principles and organizing its powers in such form, as to them shall seem most likely to effect their Safety and Happiness.[7]

As one might expect, theocracies and revolutionary republics are relatively pure types of government, clearly exhibiting their respective views of humanity. Most of the world, however, is governed by state systems that have been cobbled together over centuries. Hence, they exhibit mixes of the two assumptions. Countries like the UK and Canada, for instance, have constitutional monarchies, where vestiges of the divine right of kings reside uncomfortably with the modern, democratic superstructure. The UK monarchy initially had a divine right to rule but over the years that authority has been delimited through such measures as the Magna Carta in 1215 and the Bill of Rights in 1689. Over time the monarchy has become a constitutional system, with power to govern residing with duly elected institutions and only under clearly proscribed conditions by the sovereign.[8] While individual rights have continued to expand (such as when the UK acceded to the European Convention on Human Rights in 1951), there is no formal assertion of the power being vested in the people. Rather, the national constitutional structure continues to make self-reference to the providence of God. Canada, in contrast, had a similar experience to the UK until it began to pursue independent state making in the 1800s, culminating in the Constitution Act, 1982, where, although the preamble asserts Canada 'is founded upon the principles that recognize the supremacy of God', the fundamental freedoms and structures laid out in the rest of the Act define a humanist basis for governing, with the continued vestigial authority of the Governor General, as representative of the Monarch, to select a government that can command the majority of the legislature.

One could conceive of the issue as a 2×2 matrix of assumptions.[9] One dichotomy that is embedded in the philosophy of government is assumptions about the 'state of nature' of man.[10] Some Enlightenment philosophers, such as John Locke[11] and Jean-Jacques Rousseau,[12] have a positive view, assuming that man is inherently flawed but is perfectible. In contrast, philosophers such as Thomas Hobbes see man as fatally flawed, and life in the state of nature characterized as 'solitary, poor, nasty, brutish and short'.[13] This dichotomy distinguishes whether one is optimistic or pessimistic about the role of society and whether voluntary or coercive authorities will be needed. Meanwhile, a second dichotomy relates to whether one sees the states as the result of a social contract (a bottom-up aggregation) or a divinely justified system. This distinguishes between circumstances where the individual or collective could dominate. Putting these two dichotomies together suggests a typology of how governments may orient themselves, with an array of libertarian to hierarchical structures, with varying degrees of coercion and differing levels of risk of incompatible governing structures.

## STRUCTURES OF GOVERNING

Given that governing is 'the ability to do or affect something', the practical questions to ask are who the state involves, what does it choose to do and why, when and where does it act and how does the state go about exercising its power to get things done. The OECD offered a useful schema for thinking about how states go about getting things done, postulating a five-stage policy cycle, including: agenda setting (for national, sectoral or social strategies); design (where issues on the agenda are formulated into concrete initiatives, programs or policies); implementation (when the designed activity is operationalized, which often requires some changes); evaluation (*ex post* review of processes and outcomes); and policy learning (when lessons from the four earlier stages are absorbed and incorporated into the agenda setting exercise).[14]

Without delving into the specific activities undertaken in pursuit of completing a policy cycle – which would necessarily change as the nature and organization of state systems vary – it is possible to identify some critical features that will need to be part of any governing system. All state systems will require some authoritative principal or agent empowered to lead the process, some criteria for including or excluding issues or participants, and some process for reaching a conclusion.

Who decides in a state system is partly determined by the orientation of the state and its particular relationship with its population. Democracies will inevitably differ from theocracies, autocracies, oligarchies and various other authoritarian structures. The more individuals are viewed as the foundation for the state, or partners in a social contract, the more the state will consider the views of the general public. If fewer individuals are part of the system, then special interests (such as class, oligarchs and kin) are more likely to dominate.

At root, democratic systems only justify their existence if policy matters are decided by the preferences of the majority in some representative decision-making process, often involving elections or referenda. In that context, political scientist Robert Dahl proposes five standards for democracy: equal and effective opportunities for participation; equal voting; equal and effective opportunities for learning about policy options and implications; membership control of the agenda; and universal suffrage.[15] To deliver this, a modern, large-scale democracy would then need six political institutions: elected officials; free, fair and frequent elections; freedom of expression; alternative sources of information; free right of association; and inclusive citizenship. There are a wide range of explanations of exactly how such a system would work and whether it could actually sustain participation by citizens and deliver outcomes that improve greater good.

Economist Anthony Downs, for one, was somewhat skeptical that such a system could function. In response, he constructed a comprehensive theory of democratic decision making based on economic motivations, assuming that both voters and politicians will engage in self-serving behavior, such that electoral candidates and parties would locate themselves in the political spectrum in order to maximize their chances of being elected.[16] This median voter model asserts that successful candidates tend to take a position at the median of a normal distribution of voters, which tends to drive the democratic system to deliver median views, which at times can diverge significantly from the national average and is likely to deliver results that are not fully commensurate with the public good. Furthermore, Downs suggests that prospective voters in such a system will have limited incentives either to inform themselves about the issues or to vote.[17] Depending on the nature of the electoral system, one could conceive of the system as reflecting class, corporatist interests, special interest groups or other subsections of a society.

Regardless of how the system operates, in light of the adage that 'politics is the art of the possible', it is impossible to expect that every decision in society will be subjected to the full-scale scrutiny of majority decision making. Although some states, such as Switzerland, may be able to administer referenda on a large number of issues annually, most countries cannot practically deliver such a process. In those cases, countries use a set of representative processes, involving general elections (where the government must represent at least a plurality of voters and present themselves for periodic assessment in general elections), responsible executives (which exercise delegated authority from the legislative arm in day-to-day governing but in turn must accept responsibility for problems and failures) and various levels of parliamentary and public involvement in policy formulation (in the legislative, executive and administrative processes accordingly).

What is targeted and why those goals are targeted are inextricably linked. Some assert that what governments choose to do is fundamentally and inherently a matter of values.[18] In the first instance, a government's orientation to its citizens reflects a set of values – this can be observed in most declarations of human rights, including many national Bills of Rights and in the UN Declaration of Human Rights of 1948. Beyond that, choices may appear to be made on less prosaic grounds but they most fundamentally are driven by the values of the society and the governing system. These values would determine whom one is governing for and what sort of 'order' is appropriate for that group. Although democracies may often fall short of their goals, they usually aspire to create and maintain an order that will enable citizens to exert their liberties. Drawing on Locke,[19] many liberal, democratic governments allow citizens to make their own decisions about how to work and live and, within limits, will accept the resulting outcomes. Over time, however, there has been increasing concern about inequality. Some states are most concerned with inequalities of outcomes (including incomes, wealth, health, education or social status); others are more concerned with inequalities of opportunity (such as childhood poverty, universal education and labor mobility). Most states, however, exert their authorities in both realms, providing a variety of incentives to create opportunities while at the same time enacting differential taxes and support measures to moderate the outcomes.

Beyond the basic choice of what order to support, most governments have adopted more engaged policies. Even in a relatively libertarian world, there is a narrow economic rationale for the state to engage in the economy and society when there are public good failures (that is when private initiative would produce too little of a non-rival, non-excludable good, such as public health or national defense) or externalities (when either too little of a good, such as research, is produced or too much of a bad, such as industrial pollution, is produced because the market is unable to incorporate the spillover effects into the market price). In those cases, the state may subsidize, tax or regulate activities to improve their production and distribution. Other governing systems, with fewer democratic ideals and values, may choose to encourage or impose a different order. Many governments do not trust their citizens to know best and are unwilling to accept the distributional effects of individual choices. In those cases, they may use their power to support some other outcomes, be they to the benefit of the many (socialism), the few (oligarchies) or special interests (such as corporations or ruling elites). Alternatively, theocracies (such as the caliphate) define order in the context of spiritual outcomes. Depending on the orientation, different actors will be engaged, different subjects and issues addressed and different options considered. This is particularly relevant for new technologies, as some have the potential to strengthen power systems, others have the potential to undercut power systems and some may be inimical to the values and beliefs of rulers or the ruled.

When and where a state acts is also tied to the values and interests of the system. Although there have been no definitive studies of this, one might suggest that those states that are more democratic will tend to live in the here and now. People generally are simply too busy, disconnected or disinterested in the broader world or in the distant future to exert too much energy on the subject. This is not to say that some people and groups are not thinking beyond the present place and time. Religious groups, the development community, the environmental movement and the research community all in one way or another envisage a better world (or at least fear a deteriorating world) in the near or distant future. Each group espouses a different order. But, taken as a proportion of national or global populations, these communities are minorities and have only limited power to influence states and have only at times gained ascendancy and power to define the scope of state policy more broadly.

Perhaps the most interesting question is how the state makes choices. Chapters 9–12 will examine further specific examples of state action, but for now it is useful to examine how states in general decide. Each type of government – democracy, autocracy, oligarchy, theocracy – will have embedded in its foundational values a framework for decision making. The more power is concentrated in the hands of a few (autocracies and oligarchies), the fewer interests will be engaged and the less the options considered may conflict. Theocracies, similarly, do not rely on engaging large groups of individuals in decisions; rather, scripture and faith, interpreted by some ecclesiocracy, will divine the will of God.

Democracies present more interesting problems. Given that they have to balance individual autonomy and social order, they often are forced to develop complex processes for legitimating decisions. One simple rule is to argue that the majority should rule. The difficulty is sorting out the majority view: general elections, plebiscites, polls and other methods have been used to gauge opinion. But such approaches can offer a false promise in politics, forcing a common denominator where there often is none.[20] In response, many states seek to supplement raw numbers with more nuanced views.

Going out and seeking advice, however, still requires a set of rules. The International Association for Public Participation has defined a 'public participation spectrum' that describes five different levels of public involvement, representing differing levels of public impact. The spectrum identifies the flow of information (from government and to government), the degree of involvement and the level of empowerment, differentiating between five levels of engagement, ranging from the lowest level which involves only one-way flows of information from government to citizens to the highest level with decision making in the hands of citizens. In between there are three varying levels of public engagement where governments solicit public opinion. First, governments can consult to obtain public feedback through a range of focus groups, surveys and public meetings. Second, governments could involve interested individuals and groups through workshops or deliberate polling. Third, governments at times collaborate to build a consensus and ultimately to incorporate public advice and recommendations into decisions to the maximum extent possible.

For the intermediate levels, where information and inputs are being received from the public, it is not clear how that information is to be used in decision making. Input from a wide range of stakeholders leads to conflicting opinions and resolving different views into a decision can be challenging. For example should the government consider the views of the majority or should all actors compromise to reach a consensus? Canadian policy analysts Leslie Pal and Judith Maxwell offer a six-point framework that categorizes how governments might go about resolving conflicting input from various participants.[21] Decisions can be made or justified based on the process used; open, transparent and inclusive systems can in and of themselves justify a decision. A second option is to go with the majority opinion, determined by referenda, votes or polls. Third, a government might appeal to a utilitarian rationale, where the public interest is a balance or compromise of different interests based on intensity of reactions (in an effort to avoid tyranny of the majority, or sometimes tyranny of the loudest!). A fourth option is to appeal to and seek a consensus on common pragmatic public interests. Fifth, the state can use shared values or normative principles to decide. Although we may have multiple reference points in decision making, the concepts of universal self-interest and maximization of utility drawn from Smith and Bentham now generally dominate in our policy systems.[22] These concepts have been profoundly radical because they supplanted the prerogatives of the few with the interests of the many. This has led to what University of Toronto political scientist Janice Stein calls the 'cult of efficiency', which at times may improve outcomes but has the potential to overwhelm other legitimate objectives of the state.

## TOOLS OF GOVERNMENT

Despite the fact that a large body of scholarship and literature is devoted to the consideration of the tools of government, for the purposes of this analysis we can lump the options into three categories of tools: command and control systems; incentives and disincentives; and regulation.

States initially were conceived of and operated as command and control structures. Individuals under pre-Enlightenment systems were often viewed as chattels of the ruling system, with their labor, output and even lives at the beck and call of the rulers. Although there was some room for independent action, it was limited. Since the Enlightenment, individuals have increasingly moved to the center of many state systems. The gradual adoption of Locke's dictum that men should 'own the fruits of their labors' made it more difficult, and less certain, that direct commands and controls would be effective. Although vestiges of the command and control system remain in public enterprise, even there the ability to regulate the actions of individuals is increasingly limited. Command and control approaches can also be a poor choice when problems defy full description – governing systems are often unable to assemble the right knowledge to make effective rules.[23] Command and control is simply too inflexible, costly, cumbersome and inefficient.

In the place of command and control, states have constructed complex approaches to redirect individual action to conform to state goals. Many states use their fiscal powers and monetary authority to change the incentive systems for individual action (subsidizing actions desired by the state and taxing actions the state wants to discourage). As a result, a blended state-market approach has evolved. Although much of this effort did have an influence on individual actions, there has been a tendency for the administration of such incentive systems to become stultified and self-serving. David Osborne and Ted Gaebler sparked a renewed debate about the appropriate role for the state in the economy and society with their book entitled *Reinventing Government*. They asserted that states needed to become more 'entrepreneurial', ultimately shifting to 'systems that separate policy decisions (steering) from service delivery (rowing)'.[24] Such entrepreneurial systems would promote competition, empower citizens, focus on outcomes rather than inputs, be mission or goal directed, define their 'clients as customers', anticipate and prevent problems before they emerge, earn (not just spend) money, decentralize authority, prefer market mechanisms to bureaucratic mechanisms and seek to catalyze all sectors, not just the public domain. Osborne and Gaebler suggest, in one way, this is 'third-party government'. They posit there are 36 alternative forms of service delivery, ranging from the traditional (laws, regulation, licensing, taxing, grants/subsidies/loans/guarantees and contracting) to the innovative (franchising, partnerships, quasi-public corporations, procurement, insurance, rewards, technical assistance, information, vouchers, volunteers, jawboning) to the avant-garde (seed money, equity investments, voluntary associations, quid pro quos, demand management, sale/exchange of property and restructuring the market). They believe the focus now is on using the best sector – public, private or voluntary – to deliver services.

The one area that most scholars think is increasingly important is regulation. Political economist Giandomenico Majone argues that regulation has become the new border between the state and the economy, and the battleground for ideas on how the economy should be run.[25] This is especially true for transformative innovation, where the greatest opportunities and challenges are being faced.

## THE REGULATORY STATE

A number of scholars assert that regulation has become the preferred policy tool for many governments, prompting claims that we now live in the age of the 'regulatory state'.[26] There are at least three textbook definitions of regulation that offer concentric circles of ever widening scope: first, regulation in the narrowest sense involves promulgation of rules by government accompanied by mechanisms for monitoring and enforcement; a somewhat larger definition incorporates any direct state interventions in the economy, which would include both rules and other incentive structures put into place; and, most widely, regulation could encompass all mechanisms of social control or influence affecting all aspects of behavior from whatever source, intentional or not, which opens the field to all pacts of commission or omission.[27] The difficulty is that these textbook definitions are both too loosely focused (by including both actions and inaction and collateral effects) and too tightly centered on the state. British scholar Julia Black argues that one could perhaps more constructively investigate regulation as the 'sustained and focused attempt to alter the behavior of others according to defined standards or purposes with the intention of producing a broadly identified outcome or outcomes, which may involve mechanisms of standard-setting, information-gathering and behavior modification'.[28] Black's de-centered definition adds particular value as it captures the changing nature of regulation, which has recently expanded as the definitions, actors, forms, domain and instruments have evolved. Although many countries may exhibit some characteristics of a regulatory state, many more appear to exemplify characteristics of a regulatory society, where regulation is de-centered from the state and even from well-recognized forms of self-regulation. This de-centering has flowed from the way our world is becoming more complex, fragmented, interdependent and ungovernable and from the reality that the distinction between the public and private is vanishing. In such a world, regulation is co-produced, not formally sanctioned by the state, and seldom is confined to any specific territory. De-centered systems tend to be hybrid creations of government and non-governmental actors, multifaceted and indirect (using multiple, simultaneous measures).

Within any regulatory system, wherever its borders are drawn, there are two main alternative ways that regulations might be structured.[29] The normative view (espoused predominantly in the US) asserts that public regulation of economic activity is justified only when the market is incapable of producing a social (Pareto) optimum. Regulatory action is thus justified when there is some sort of market failure: monopoly; externalities; incomplete or asymmetric information; or public goods. This narrow view of regulation would suggest that the system operates in the best interests of the majority, using some majoritarian or utilitarian method of development. The contrasting, positive theory, formulated by George Stigler and others, suggests that self-serving individuals and groups will capture a regulatory system, forgoing the public interest (defined as the protection and benefit of the public at large or some significant subclass) and instead design and operate the system primarily for the benefit of some small, privileged subgroup in society.

At a practical level, the legal systems that underpin and give effect to regulatory systems are one major dividing line in the world today. The lawyers at University of Ottawa posit that there are four main legal traditions: civil code, common law, religious law and customary law.[30] Talmudic and Muslim law (autonomous legal systems which are religious in nature) and customary law (rooted in wisdom born of concrete daily experience or more intellectually based on spiritual or philosophical traditions) are important influences in some countries, but most countries have also adopted some or all of the traditions of civil or common law. As of 2002, an estimated 87 countries representing 33 per cent of the world's GDP used the civil law system, 43 countries representing 27 per cent of the world's GDP used common law and only a handful of countries used religious or customary law (accounting for less than 0.5 per cent of world GDP). The rest of the world, more than 60 jurisdictions representing about 40 per cent of global GDP, used some mixed system involving one or more of the two main systems.

Civil law is usually defined as the legal tradition which began in AD 528–34, when Emperor Justinian commissioned the codification of Roman law in the *Corpus Juris Civilis*, and further developed in Europe and elsewhere. The modern form of civil law used in most countries today can be traced to the French Civil Code of 1804. All of Europe (except the UK), all of Latin America and most of Africa and Asia (except the former British colonies) rely wholly or partly on civil law. The main inspiration from the Roman law heritage is that precedence is given to written law; as such, most countries opt for a systematic codification of their general laws.

The common law legal tradition, which began to evolve in England in the eleventh century, is based more on case law – that is judgments on individual contests – rather than systematic legislation. Like civil law, the common law system has taken on a variety of forms. Many countries have some codes, legislation and other legal instruments, but rely on common law jurisprudence for the character of their legal system. Most of the countries that use the common law system have, at some time, had some connection to the British Empire.

The two systems have different effects.[31] Civil codes and statutes are concise whereas common law statutes are precise. Civil law statutes provide no definitions but rather state principles in broad, general phrases; common law statutes provide detailed definitions and each specific rule sets out lengthy applications or exceptions. This difference in style is related to the function of the statutes. The principles incorporated in civil law statutes need to be stated concisely if the code is to be exhaustive; common law statutes must be precise because common law courts restrict rules to the specific facts they are intended to cover. This difference has the potential to make it easier to draft common law rules, as they are not intended to be comprehensive, but rather are to simply deal with a specific matter.

We can see the effect of this difference in the way different countries regulate new technologies, especially biotechnology regulation in the US and the EU. Abstracting from any potential effects and intents of the two resulting regulatory approaches, one can observe that the regulatory structures in the US, a common law system, were cobbled together from an array of existing legal authorities and have evolved in small, incremental steps over the past 20 years. In contrast, the EU system was installed in two discrete steps (1990 and 2001) following extensive debate and consideration of the entire system, and with the introduction of comprehensive, purpose-built regulations.

The two systems also operate in different ways. The civil code tends to argue disputes in the context of the rights and obligations encoded in the principles of specific statutes whereas common law disputes focus on fact-patterns. Common law advocates analyze cases presenting similar but not identical facts, extracting from the specific rules, and then, through deduction, determine the often very narrow scope of each rule, and sometimes propose new rules to cover facts that have not yet presented themselves. In a perfect world, the legal effect of living under civil code or common law systems may be minor but, in practice, the impact can be significant. Civil code systems, though comprehensive, often have difficulty resolving cases that do not fit clearly within existing statutes; it is both possible and likely that at times cases that present themselves may not be able to be adjudicated. In contrast, the common law focus on case-by-case assessment and on finding remedies ('remedies precede rights'), with the result that in every situation some sort of a decision can be made.

Although the differences in theory between the two systems may be small, the practical effect is more significant. One simply needs to try living in the two regimes to get a sense of their differential impact. People living in both jurisdictions are on paper more or less equally regulated but individuals and companies seeking to operate in civil code systems tend to find a larger number of inclusive rules and processes that govern parts of their lives or operation, with a greater degree of variance in the administration of those rules than in common law systems. This leads to a tendency within civil code systems for economic and social development to await the development of permissive legislation and regulation. In common law systems, in contrast, proponents of change are generally free to do whatever is not explicitly prohibited. There is risk in both systems that what is possible may be prohibited, but the common law places the onus on those wanting to stop a development to make their case to a greater extent than does the civil code. Ultimately, the concise nature of civil code systems has the potential to slow the process of change, which can work against the forces promoting transformative change.

## CONCLUSION

Understanding the role for the state is crucial to understanding how we can manage transformative innovation. The state has a unique role that only it is able to deal with – policy making without politics, or alternatively a public choice (market) framing of governance, would neglect certain missions of government. Political scientist Ali Kazancigil asserts that 'the market cannot cope with civic issues and citizenship, but only with customers/consumers. Democratic politics is about the representation and mediation of general interests, the market concerns only exchange and negotiation between sectoral interests.'[32]

The role of the state, by stalling, unleashing, or leading technological innovation, is a decisive actor in the governing process. Only it can organize and express social and cultural preferences.[33] In pursuit of governing technology, the state engages in a mix of encouragement, information gathering and dissemination, risk evaluation, economic and social management and conflict resolution, alternatively acting as proprietor, promoter and regulator.[34]

Two countervailing trends have put the state in a difficult position. On the one hand, citizens have an increasingly higher set of expectations about their governments. This is reflected in the wide-ranging demands for audits, investigations and commissions of inquiry into the actions and performance of our heads of state (such as Tony Blair and George W. Bush), our administrations (such as Human Resources Development Canada) and our regulators (including congressional and parliamentary reviews of public health regulators in

Canada, the EU and the US). One might argue they have reached a state of hyper-vigilance that demands more and better technical accountability, responsibility and transparency (the hallmarks of modernity). Meanwhile, increasing complexity and heightened awareness of new innovations, especially new technologies, is raising demands for more inclusive, values-based decision making. Although we have some emerging processes for handling such processes,[35] this type of a post-modern system is grounded on completely different expectations and principles. Reconciling the two trends will necessarily require us to more clearly understand both the competence and capacity of government, but also the respective roles of the market and other civil authorities.

## NOTES

1. Homer-Dixon (2001: 150).
2. Hobbes (1651).
3. Banacek (2005).
4. Ruttan (2003: 111), cites Eckstein's test of theories of political development as a basis for this assertion.
5. The constitutional peasant skit in the movie 'Monty Python and the Holy Grail' makes this point with humor. The peasant named Dennis challenges Arthur's claim to be 'King of the Britons', asserting: 'Listen. Strange women lying in ponds distributing swords is no basis for a system of government. Supreme executive power derives from a mandate from the masses, not from some farcical aquatic ceremony . . . You can't expect to wield supreme executive power just 'cause some watery tart threw a sword at you . . . I mean, if I went 'round saying I was an emperor just because some moistened bint had lobbed a scimitar at me, they'd put me away!' At one level, all processes for justifying power involve some similarly farcical justification.
6. Dictatorships are a little different, but ultimately parse into one of the two basic types. Although dictatorships often begin as revolutionary movements, once a regime is established, generally the rulers adopt one of two founding myths: if God is not invoked directly, often autocratic rulers try to define themselves in mythical, god-like terms; alternatively, many autocratic systems will use their role in the revolution to justify themselves as a reflection of the will of the people. Thus, ultimate power is vested either with god or the individual.
7. China represents a good example of how a revolutionary autocracy justifies its authority. The preamble to the Constitution of the People's Republic of China, 4 December 1982, offers an essay on the historical suppression of the people under rule of imperialism, feudalism and bureaucratic capitalism, and justifies socialism. Article 1 asserts that the PRC is a 'socialist state under the people's democratic dictatorship led by the working class and based on the alliance of workers and peasants'. In short, it justifies an autocratic regime based on representing the will of the people.
8. Bagehot (1867).
9. 6 (2001).
10. The difficulty with any consideration of the state of nature is that we cannot deductively test our assumptions. A few have suggested that we might be able to get a hint about whether we are perfectible or damned by looking at groups of men placed outside the normal bounds of society (for example, in prisons). Viktor Frankl in *Man's Search for Meaning* (1962) assessed human responses to the dehumanizing Nazi death camps in the Second World War, concluding that humans seek meaning in positive relationships, even in situations of abject

horror. The difficulty of taking this as a reflection of man's underlying nature is that most of the victims in those camps shared a common religious base, which clearly would not exist in a true state of nature. Hence, we are left with making untestable assumptions.

11. Locke (1689).
12. Rousseau (1762).
13. Hobbes (1651).
14. OECD (2005).
15. Dahl (1998).
16. Downs (1957).
17. Hardin (1999).
18. Stein (2001).
19. Locke (1689).
20. Stein (2001).
21. Pal and Maxwell (2003).
22. Stein (2001).
23. Gunningham and Graboksy (1999).
24. Osborne and Gaebler (1992: 35).
25. Majone (1994).
26. For example, Scott (2002).
27. Black (2002).
28. Black (2002: 20).
29. Majone (1994).
30. University of Otttwa, Faculty of Law (No date) World Legal Systems. Accessed on 25 October 2005 at http://www.droitcivil.uottawa.ca/world-legal-systems/eng-monde.html.
31. Tetley (2000).
32. Kazancigil (1998: 72).
33. Castells (1996: 13).
34. 6 (2001) and Phillips and Khachatourians (2001).
35. For example, IAPT2 (2000); Pal and Maxwell (2003).

# 7. Governing through the market

> One of the great intellectual achievements of modern economics has been to work out very precisely the circumstances under which decentralized systems of market exchange can produce results that are efficient . . . much of the pattern of modern life has emerged without ever having been consciously willed by anyone.
>
> **Paul Seabright**[1]

## INTRODUCTION

The brief discussion in Chapter 1 that recounted some of the many uses and changes of the lowly potato offers a starting point for understanding how both old and new technologies are governed through the marketplace. Generally markets require three things: voluntary buyers; voluntary sellers; and some form of price discovery to match potential supply and demand. New technologies at times have none of those opportunities. These factors in many cases need to be created, adapted and adopted, all of which have both economic and non-economic aspects.

Different technologies will have different needs and different responses. Iterative technologies, such as those that simply allow for cheaper production, often are relatively easy to fit into existing production and marketing systems. Similarly, small, incremental improvements in the quality of the product, which do not introduce any new attributes but simply differentiate the product based on enhancing existing attributes, would face few challenges in entering and sustaining a market. In most of these cases firms will have benchmarks against which to align their technologies and to establish competitive pricing structures. For example, the introduction of new cultivars in the potato industry is a relatively seamless process. Individual cultivar developers are able to compare their new varieties against the best of the rest, set a competitive price, adapt the cultivars to fit with the current production and marketing systems and search out potential buyers. For the most part, the supply channels are easily adapted to new varieties and consumers are generally indifferent to most small, incremental changes. This process of continual, iterative improvement in most technologies happens almost without need for new institutions or settings. That is not to say that non-economic structures are not important, but rather that the prevailing institutions are able to manage the job. It is also important to keep in mind that the vast majority of all inventions are never successfully commercialized (by some estimates, fewer than 10 per cent of patents have commercial value and less than 1 per cent are of seminal importance).[2] Their failure, one might argue, is not due to a lack of institutional support, but more likely reflects the limited value offered by the new idea relative to the significant fixed costs all new technologies and products face when entering the market.

Novel, discrete innovations that create new product categories or new technologies that require major changes in the production or marketing channels are more problematic. The VHS player, for example, created a whole new category of entertainment for consumers. It had neither existing production or marketing channels nor a readily identifiable niche in the consumer market. In response, production and marketing teams had to work through either the open market or some non-market process to construct the institutions that would create both supply and demand. Although there are some examples which can offer some insights, the uncertainty, time and costs of such undertakings are inevitably going to be exponentially higher than the costs of introducing an iterative innovation.

Those technologies that are fundamentally transformative face the greatest challenges. In most cases, the costs of researching, developing and commercializing a transformative technology are going to exceed the returns available in any single market – the return on the investment is going to be realized only with a wide range of uses. This poses a number of challenges that the market is often unable to manage. Ensuring an appropriate supply is no longer a matter of fitting into an existing supply chain, or even of creating a new channel – it will involve an array of adaptations and innovations across a wide range of production systems. The more markets that are involved, the more difficult it will be to manage the supply response. Similarly, the demand side is often intractable. Although individual uses of a transformative technology may be easily explained and marketed to some consumers, the overarching and unknowable nature of transformative technologies means that unexpected or unintended effects in one market can have an impact on consumer attitudes in other markets. The *Bt* potato, to some extent, suffered from the knock-on effect of controversy both about the regulation of food and public health (following a series of food safety and medical therapy scares in the 1990s) and potential environmental and social effects in other products and in other markets. The uncertainty (in the sense that there are no credible probabilities of possible outcomes) surrounding transformative innovations makes them potentially high-return investments but also makes them very high-risk opportunities. Quite simply, markets often are incapable of dealing with uncertainty without some outside assistance.

Ultimately, the challenge is to understand why some markets function and others do not.

## MARKET ASSUMPTIONS

Generally, markets are assumed to be driven by self-interests. Individuals, operating as producers or consumers, sell or buy labor, other inputs, goods or services based on their own calculations of what is in their own best interests, mediated by market prices of the various factors and goods. These markets are driven by voluntary transactions.[3] Unlike in the state or in social groups, there is no formal acknowledgment in markets that the interests of others are relevant or important; nevertheless, the economic theory of markets suggests that, under certain rigorous conditions, if everyone acts in a self-interested way, then we can optimize our economic activities. Economist Paul Seabright notes that the cooperation that delivers existing or new products often appears effortless because people and institutions embody a kind of tunnel vision, by 'play(ing) one's part in the great complex enterprise of creating the prosperity of a modern society without knowing or necessarily caring very much about the overall outcome'.[4]

Markets have both allocative and creative roles.[5] Allocation is a static concept that addresses the optimization of the mobilization of inputs and distribution of outputs in a specified period of time. Creation, in contrast, is a dynamic concept that involves incentives for changing the current optimal distributions in order to expand the production and consumption possibilities in the future. Systems that can handle static distribution may not be able to deal with the dynamic challenge. Although prices and non-price factors – including infrastructure underpinning the price mechanism, physical structures and organizational capacity of governments – are important for both roles, they pose a particular challenge for the creative activity. Generally when markets are absent or underdeveloped, markets cannot be expected to spontaneously create or generate the missing mechanisms.[6]

Nevertheless, the concept of self-directing markets has dominated much of the discussion since 1776 when Adam Smith published his seminal book, *An Inquiry into the Nature and Causes of the Wealth of Nations*. This revolutionary document laid out the underpinnings of a liberal, individualistic-based economy, which ultimately helped to redraw the boundary between the state and the market. Smith did this quite deliberately and his supporters, promoters and admirers similarly had an overt, political objective in promoting the concept of self-regulating markets as a way to govern the economy.[7] Prior to this, the feudal system dominated in many areas, with individuals having little or no influence on their role in the production or consumption systems. While there had been some entrepreneurial development prior to this, the political system continued to constrain the role of the merchant class. By positing that self-interested individuals could lead to social and economic improvement, Smith provided another underpinning to the emerging liberal conception of government based on a social contract.

Smith laid the foundations of an economy based on self-interests mediated by prices while others added the specific concepts that now comprise our understanding of the economic system. Over the succeeding 115 years, the theory was fleshed out and added to by a number of economic and social philosophers. Jeremy Bentham introduced the concept of utility and the notion of an economic calculus that could generate an optimal economy (based on a balance of individual valuations of pleasure and pain).[8] William Stanley Jevons fleshed out the marginal utility theory of value.[9] John Stuart Mill expanded on Bentham's concept of utilitarianism and added to our understanding of the economic underpinnings of bilateral exchange and trade based on comparative and not simply absolute advantages.[10] Leon Walras elaborated on the concept of a general equilibrium and introduced the simplifying mechanism of the 'Walrasian auctioneer' to effect equilibrium prices and quantities.[11] Ultimately, Alfred Marshall put it all together in a mathematical, graphical package, thereby spawning what has come to be known as the 'Marshallian scissors' of supply and demand that inexorably drive markets to an 'optimal' output of goods produced and consumed, at one single set of market clearing prices.[12] Although we have elaborated on and empirically tested the validity of the underlying theories of supply and demand over the past century (using an ever increasing array of mathematical, empirical and computer tools), the basic theory of markets remains much as elaborated by the great thinkers of that period.

Essentially, economic theory posits that individuals as producers, left to their own devices, will apportion their labor (either directly in own production or through selling their time to others for a wage) up to the point where the 'pain' inflicted by the last unit of work equals the 'pleasure' that could be derived from the output or wages one generates with that unit of labor. Similarly, individuals as consumers will buy increasing volumes of goods or services as long as the 'pleasure' (or utility) thus generated is greater than or equal to the cost of buying those goods. Individuals following this set of very basic, individualistic rules would thereby optimally allocate their own labor and income. Theoreticians have demonstrated that it is possible that an economy made up of such self-satisfying individuals could produce and consume an optimal amount of economic activity. This optimum is defined as the point at which maximum utility is reached – working longer hours to consume more would be a loss in overall utility because the pain of the last unit of work would exceed the pleasure of the last unit of goods consumed while working fewer hours would lower the pain but would lead to disproportionately larger declines in utility from consuming goods.

The production side of the theory was subsequently extended (by Jacob Viner and others) to convert the individual labor choice into a firm's cost structure. The theory suggests that in most circumstances the same qualities and quantities of one input (land, labor or capital), holding other inputs constant, will produce a smaller output for each additional unit of input. For example, adding more labor to a farm of a fixed size will eventually cause congestion and relatively lower output per each additional worker. This concept of declining marginal productivity leads to each incremental unit of output costing more to produce. Graphed in a two-dimensional space, with volumes along the horizontal or X axis and marginal value of product (usually converted to price) along the vertical or Y axis, the supply curve for any product by any producer is generally assumed to be upwardly sloping, as declining productivity of additional inputs leads to higher costs of producing a higher quantity of a good. The industry or product supply curve would simply be the sum of supplies offered by each of the producers at each price. The actual slope of the line is a function of the rate of declining marginal productivity – the faster the decline, the steeper the curve, and vice versa.

The consumption side of the economy was similarly fleshed out. The theory suggests that individuals will realize correspondingly less pleasure or utility from successively larger amounts of any good consumed. This declining marginal utility theory suggests that a consumer would likely pay a higher price for the first unit of a good they consume, but as successively more units of the good are consumed, pleasure or utility would decline and the price they would pay would correspondingly drop. In theory, at some point no further pleasure or utility would be generated by further consumption. Graphed in a two-dimensional space, with volumes along the horizontal or X axis and marginal utility (usually converted to price) along the vertical or Y axis, the demand curve for any product is generally assumed to be downward sloping, as declining marginal utility would lead a consumer to be willing to pay less for additional units. The total demand curve would simply be the sum of all of the volumes demanded by individuals at each price level. The slope of the demand curve would depend on tastes, preferences, incomes, and the prices of substitute and complement offerings.

Putting the supply and demand curves into the same two-dimensional space yields an X-like picture (many have called the figure the 'Marshallian scissors'), with an upwardly sloping supply curve intersecting the downwardly sloping demand curve at some unique pairing of a price and a volume. The intersection point, or equilibrium, is optimal because no other points will lead to as much aggregate utility. If more output is produced than the optimum, consumers would only pay a price below that which the producers' last unit cost to produce. Thus, although consumer utility could be higher, producers would be worse off. If a market entered this area, the lower prices would drive down production to the point where supply again equaled demand. Alternatively, if a volume less than the optimum were produced, consumers would be willing to pay a price higher than the last unit cost to produce, leading to unusual profits for producers. Assuming there are no barriers to producing more, production would rise until the market clearing price equaled the cost of the last unit. The process of finding a market clearing price is not absolutely clear. Walras offered the concept of a conceptual auctioneer to manage the *tatonnement* process of groping toward the market clearing price – essentially, the auctioneer posts prices, seeking bids and offers, and only allows trade once a market clearing volume and price combination is realized.[13] The operation of the Marshallian scissors and the Walrasian auctioneer thus creates a self-correcting optimum. If one accepts this construct, then there is little or no role for other governing systems – states and collective bodies could not improve on the choices of individuals.[14] Paul Seabright argues that 'one of the great intellectual achievements of modern economics has been to work out very precisely the circumstances under which decentralized systems of market exchange can produce results that are efficient, in the sense of improving the condition of every individual as far as possible whenever this can be done without harming someone else . . . the pattern of modern life has emerged without ever having been consciously willed by anyone.'[15]

Up to this point the economic paradigm is comparable to the political science paradigm, in that most of the assumptions have been derived in an inductive way. About the end of the First World War the professions diverged as economics formally adopted the deductive, reductionist scientific method. Increasingly, the study of economics moved to positivist, deductive testing of the theoretical hypotheses underlying supply and demand curves and other economic concepts. In the process, economists have had to simplify their analyses, assuming 'all else being equal' (*ceteris paribus*), including virtually all non-economic variables. Thus, economics diverged from the other social sciences, increasingly examining the isolated operations of the market and its actors. Although this has led to a greater degree of specificity of the economic relationships in society, it has worked to downplay the vital non-market functions in the economy.

One area that was often overlooked in the canon was the role of the underlying assumptions. The optimal outcome, called 'perfect competition', is more of a theoretical construct than an observable reality. Theory clearly shows that it can only be realized if five rigorous and highly unlikely conditions hold. First, all actors must be atomistic – there must be a large number of small producers and consumers in a given market, each so small that their actions have no significant impact on others (that is both buyers and firms must be price takers and not exert any market power). Second, the goods and services must be homogeneous or perfect substitutes. Third, there must be perfect and complete information – all firms and consumers must have complete information on production technologies, attributes of a product, prices and quantities. Fourth, there must be no barriers to entry or exit for either producers or consumers. Fifth, there must be no external effects of any production or consumption that lie outside the market – both positive externalities (such as spillovers from research or general health) and negative externalities (including pollution and congestion) must be known and included in the price of the product.

As different sets of assumptions are relaxed or break down, different types of imperfections arise. Monopoly and oligopoly, where one seller or a small number of sellers respectively exert market power, occur when the first (atomistic) and fourth (entry) assumptions fail (sometimes caused by the failure of the third, perfect information, assumption). We tend to see this in markets where there are physical, economic or legal barriers to entry (such as monopoly resource grants, large fixed costs or patent grants). Monopsony and oligopsony, where there is only one or at most a small number of buyers who exert market power, similarly occurs when the first and fourth assumptions fail. Monopolistic competition, where there are many sellers producing highly differentiated goods, occurs when the second (homogeneous product) assumption is negated.

When assumptions three (perfect information) and five (no externalities) fail, we face a set of market failures that are related to the inability to fully and efficiently transact to exchange goods and services. Nobel Prize winner Ronald Coase argues that, with no transaction costs and full information, private bargaining should be able to handle any and all externalities.[16] But transactions – when a good or service is transferred across a technologically separable interface, such as where one stage of processing or assembly activity terminates and another begins – are seldom costless.[17] Transactions involve three cost components – search costs resulting from the efforts required to locate individuals with reciprocal interests; negotiation costs relating to efforts to conclude an agreement; and enforcement costs incurred in monitoring the performance of the parties to the transaction – that can be classified as either *ex ante* or *ex post* costs, according to the contract's date of agreement.[18] Furthermore, transactions may differ from one another depending on the extent of communication or strategic behavior, the frequency of a transaction and the degree to which any asset involved in a transaction has only specific uses.[19] The higher the transaction costs, the less likely we will see a perfectly competitive market outcome.

Even if a competitive marketplace might equalize private social benefits and costs, if there are externalities then the marginal social benefit might not equal the marginal social cost, creating a market failure that is often characterized as a public good, a common pool or a technical externality problem. Despite overall benefits from public goods, markets fail to produce adequate amounts as firms cannot charge users the full marginal cost because no one can be excluded from their consumption. Common pool resources (such as fisheries, clean air or market development investments) suffer from the lack of exclusion and overuse can result in the depredation of the resource. Finally, technical externalities (such as pollution or knowledge) occur when activities of an individual can affect the welfare of others outside the transaction. Technical externalities, similarly, can generate imperfect competition in markets if, for example, buyers or sellers lack information about the qualities, quantities and prices on offer.

A variety of market-based institutions are often established to address these potential market failures. Common ownership (often through vertical integration) is a common response to asset specificity; alternatively, some potentially opportunistic party could make *ex ante* credible commitments (often in the form of equity infusions or long-term contracts) in order to conclude transactions that are subject to significant asset specificity.[20] Long-term contracting also can be used to address problems of opportunism.[21] Sometimes, however, market failure cannot be addressed by private action. At times governments can play the role of supra-firms, where their actions may reduce transactions costs; alternatively, collective institutions (sometimes referred to as non-government organizations, or the participation sector) may be more effective than government or the private sector.[22] Douglass North asserts that a variety of innovations related to transactions were critical to the emergence of the modern economy.[23] These innovations included increasing the mobility of capital (via the replacement of usury laws, the evolution of new financial instruments, such as the bill of exchange, and development of accounting systems to monitor agents), lowering information costs (through evolution of printed prices, manuals on weights, measures, customs and brokerage fees and integrated, efficient postal and telegraphic systems) and spreading risk (via development of insurance, joint stock ventures and portfolio diversification). Although many of these measures evolved though private initiative without any explicit government involvement, their transformative influence was only realized once the state entered into enforcement of commercial law. Regardless of the approach, however, each of these institutions (large, vertically managed firms, long-term contracts and various forms of government or non-governmental authorities) is usually governed using mechanisms other than the market. Supply and demand conditions do not operate effectively in these alternate configurations, with the result that prices cannot perform their allocative function.

Thus, even though the liberal, economic framing of markets dominates in many people's minds (and underlies the assertion that liberal markets are the best possible configuration for individual liberty), in practice the market has non-market, non-price mechanisms in major parts of its structure.

## THE ROLE OF THE FIRM

One relatively weak aspect of the economic theory of markets is the concept of the firm. Although theory has laid out a succinct, elegant characterization of supply based on a set of interrelated cost curves, it has been relatively silent on the nature of those firms that actually manage and direct the productive side of the economy. Until Ronald Coase introduced the term 'firm' in a seminar article in 1937, economics generally treated the producing side of the economy as a collection of entrepreneurs, self-employed individuals or at most owner-operators of very small, atomistic firms. These producers were believed to be governed by a combination of consumer utility theory and cost curves derived from the theory of the marginal value of product. The assumption was that all exchanges and decisions in a production chain could be handled through arm's-length, spot-marketed trading. Coase disagreed, postulating that firms can and do exist to manage the risks of contracting – whenever the costs of transacting are greater than the costs of using a non-market allocation system, firms will emerge.[24]

A large portion of the economic activity in most countries is under the control and direction of firms using non-price mechanisms to allocate internal resources. We currently have a wide range of ownership structures. Family-owned enterprise is probably the oldest and most common model. In the US, an estimated 90 per cent of all companies are family-owned and that portion is likely higher in many less developed countries. Although many family-owned enterprises are very small (often only involving the owner and perhaps a few family members), between 35 and 45 per cent of the 500 largest firms in the world are family-directed enterprises: the Ford family owns 40 per cent of the eponymous automotive company; the Walton family owns 38 per cent of Wal-Mart; the Lee family owns 22 per cent of Samsung; the Koo and Huh families own 59 per cent of Hyundai; the Agnelli family owns 30 per cent of Fiat and 100 per cent of IFI; the Cargill-MacMillan family owns 8 per cent of Cargill; the Quandt family owns 47 per cent of BMW; and 42 per cent of PSA Peugeot Citroën is family-owned.[25]

One of the social innovations often credited with precipitating the first industrial revolution and contributing to succeeding rounds of innovation is the limited liability, joint-stock company.[26] Initially an innovation in Britain in the 1700s (and credited with raising the capital to support development of the early colonies in North America), the model has been adopted globally (known as an *Aktiengesellschaft* or AG in Germanic regions, a *societe anonyme* or SA in French jurisdictions, a *sociedad anonima* or SA in Spanish regions, a *Naamloze Vennootschap* or NV in Dutch countries or *Societa per Azioni* or S.p.A in Italian areas).

Joint-stock companies pose unique challenges to the economic theory of firms. A joint-stock company is a type of limited liability partnership where capital stock is issued to 'shareholders' who are empowered to engage in the management of the firm (in proportion to their share of the capital) and are able to transfer their shares to others without needing consent of the other shareholders (traditional partnerships have specific barriers to entry and exit). The challenge is that many joint-stock companies have come to have such widely dispersed ownership that there often is no single dominant shareholder or small group of highly interested shareholders who will assume leadership for the enterprise. Although there are no readily available statistics on the size of the sector, we do know that the key bourses in the G7 nations in 2005 combined had more than 22 000 joint-stock companies listed on their key exchanges (there are likely to be many more joint-stock companies as this does not include over-the-counter trades and listings on smaller, regional markets). The absolute number of these firms is insignificant compared to the estimated 300 million firms operating globally but their scope of operations is large. Comparing the market capitalization of all domestically listed US companies to the US GDP suggests that for every dollar of GDP there is about US$1.30 of outstanding capital engaged in joint-stock companies. Furthermore, large joint-stock companies employ on average more than half of the non-farm labor in most jurisdictions, and for the most part generate higher productivity and a larger share of total output than other forms of organization.

Although these companies face external market pressures (mainly supply and demand conditions in input and output markets) and are constrained both by their charters and national regulations, their day-to-day management is for the most part based on non-market principles. Most of the companies are now managed by professional managers who are supervised by boards of directors who personally own only a tiny share of the total capital base. In absence of either direct market control or clearly delineated internal governing rules, these enterprises often settle on non-market rules and mechanisms for allocating resources internally.

There is almost a limitless supply of explanations of how professional managers operate. Although the roots of modern management theory can be traced to classical economics,[27] the organizational theories of Marx and Weber and the practical advances in quality control of Babbage, the first explicit management theories emerged only about a century ago. Frederick Winslow Taylor in 1911 produced his classic *Principles of Scientific Management*, asserting that scientific management, relying on time and motion study, was one of the best ways 'to extract and codify in advance the exact attributes of both the production process and final product'.[28] That approach was adapted and adopted by Henry Ford to the automotive assembly lines in Detroit, MI. The resulting Fordist revolution kick-started a wave of institutional innovation in heavy industry that led to the development of the modern industrial economy.

Since then there has been a wide diffusion of management philosophies adhered to by various business leaders. One website offers a veritable smorgasbord of more than 250 current models, theories and methods, categorized into five streams: strategic value creation; valuation and decision making; organizational change and culture; communications and marketing; or leadership and management.[29] One can argue that in practice three distinct perspectives on the strategic management of the firm are current in joint-stock companies: the product-based view; the resource-based view; and the knowledge-based view.

Much of the earliest management literature focused on the external elements that affect firm performance. This stream of the literature is known as the product-based view of the firm.[30] The strategic objective in the PBV is to dominate in a product or service market. Firms are able to dominate when they have the best understanding of the industry, particularly the opportunities and threats that are emerging. This perspective is demand focused; firms are challenged to identify changes in consumer tastes and preferences and respond to exogenous economic shocks. Only after a firm has analyzed the competitive environment are its resources selected and capabilities developed. The PBV of strategy asks what it is that the firm must do to dominate in its selected industry. Pursuit of these answers is expected to generate the most desirable internal configuration.

The resource-based view (RBV) focuses on the internal elements of a firm. In the resource-based perspective a firm first looks inside to answer the question 'does the firm have what it takes to be number one?' This perspective recognizes that firms evolve over time and that there may be times within a firm's lifecycle that are better suited to pursuing dominant market positions (in contrast, the PBV identifies dominance as the only desirable competitive position); other goals may be relevant at other times in the RBV. The underlying principle in the resource-based view was introduced by Edith Penrose, who described the firm as 'an evolving collection of resources' that over time will change because of their ability to learn or acquire new resources and capabilities.[31] The strategic goal, from the perspective of the RBV, is to secure the factors required to build core competencies.

Both the PBV and the RBV recognize the importance of, respectively, internal and external firm elements. The PBV segregates the firm into value activities and value chains that collectively build a competitive position.[32] The

RBV recognizes that internal resources and competencies are meaningless without external opportunities and threats to respond to.[33] As the literature has evolved the contrast between internal and external firm perspectives has appeared as divisive. Upon closer inspection these two perspectives are not competing to explain the same phenomenon but rather they describe strategies for achieving two different kinds of innovativeness. The product-based view operates reactively and generates demand-pull, outcome-based, incremental innovations. The resource-based view operates proactively to generate supply-push, process-based, novel innovation. The resource-based view of the firm implies, thus, that firms pursue the monopolistic advantages of disequilibrium through a process of Schumpeterian competition, path dependencies, first-mover advantages and irreversible commitments and use of complementary or co-specialized resources.[34]

A growing number of researchers and practitioners view knowledge as the most important productive resource in determining market value and the key source of Ricardian rents.[35] As its name implies, the knowledge-based view (KBV) of the firm focuses on the knowledge attributes of a firm, which might be viewed as a special case of the RBV. The strategic objective is to create and apply knowledge in such a way as to establish or sustain competitive advantages. Knowledge forms the basis of a firm's core competencies, which in turn generate competitive advantage. The objective of a knowledge-based strategy is to identify knowledge-based resources and to maximize the value that can be derived from the deployment of those resources. Firms competing from a knowledge-based strategy rely on creating, protecting and applying specialized knowledge to the production of goods and services. The strategy can involve firm or industry specific rules and directives, sequencing, routines, group problem solving and decision making.[36] The successful use of these integration mechanisms requires that the individuals participating have some form of common knowledge, built upon language, symbolic communication (such as literacy and numeracy), specialized concepts (jargon), shared meaning (metaphor, analogy and storytelling) and a recognition of individual knowledge domains.

Although firms are fundamentally economic actors, they clearly engage in a variety of non-economic allocation systems. Depending on which theory they pursue – PBV, RBV or KBV – they will adopt internal management systems that focus on managing different sets of assets, different resources and different knowledge. Given the difficulty of determining supply and demand inside an enterprise, price mechanisms seldom are available to allocate resources. Instead, a variety of purpose-built structures or hybrids of market and non-market systems are used which may at times emulate the allocations markets might conceivably make and at other times diverge significantly from what economic theory might prescribe.

## MARKET MANAGEMENT OF INNOVATION

Economists and firms have accepted that knowledge is different from other traditional products. The cost structure of knowledge, combined with the non-rival and often non-excludable nature of knowledge, make it difficult for the unaided market to deliver the optimal amount of research, development and commercialization of new knowledge.

Any good, service, technology or product will exhibit a different range of two key characteristics: rivalry and excludability. Rival innovations result in a good or service that can only be used by one person at one time (such as a consumer durable or personal service). Non-rival innovations involve an output (usually knowledge in the form of a recipe or formula for doing something) that for little relative expense, or in some cases no cost, can be disseminated to and used by every producer in a country or the world, and no one's use is limited by other's use. Excludability (sometimes referred to as separability) measures whether the innovation is protected from widespread use by legal means (such as with a patent) or whether its adoption is limited by industrial organization requirements or climate. The more disembodied an invention, such as a recipe or blueprints for undertaking some process (as well as the creative component embodied in music, art, drama and writing), the lower the congestion or rivalry and the more difficult and expensive it is to exclude others. At the extreme, new ideas approach public goods. If a creation or invention is excludable (by some physical or legal way), then a creator or innovator should be able to appropriate a large share of the benefits from the invention, thereby justifying their investment of time and resources in discovering and developing the idea. If it is not excludable, then an innovator is unlikely to get much of a return on any investment in a new idea. This problem is being played out in a number of sectors currently. In the music industry, for instance, the music that is played on iPods costs a lot of money to develop and market. That industry has been jeopardized by the increasingly cheap and easy ways people have of copying or pirating the music (through dubbing to tape or downloading MP3s from the internet). Whenever someone does that, they don't pay any of the development costs.

The underlying cost structures differ significantly depending on the nature of the product. To illustrate the point, let's compare a traditional product with knowledge. Although traditional products can face highly variable fixed costs, they almost always face rising marginal costs (due to decreasing marginal productivity). Thus, although the firm or industry will initially face downwardly sloping fixed and average cost curves (as the fixed and total costs are amortized over increasing units), at some point the rising marginal cost per unit will exceed the declining fixed cost per unit and the average cost per unit would begin to rise. Economic theory posits that under conditions of perfect competition, with free entry and exit to this industry, firms would produce at the point where the marginal cost equals the price, which will also be where the long-run, average-cost curve is at its lowest point. At this point firms are fully compensated for all of their fixed costs but there are no extra profits. In this way, industry is self-sustaining as it gets just enough revenue from the market to justify its investments.

In contrast, the generation of knowledge requires high fixed capital investments early in the research, development and commercialization stage while in the adoption and knowledge transfer stage there are often low to negligible marginal costs to transfer new disembodied knowledge. Two demand conditions could hold. If there was one buyer who would benefit from and therefore could pay the full cost of developing the technology or idea, then there should not be a problem. Difficulty arises when there are lots of customers but none derives enough benefit from the invention to buy it outright. At that point, it becomes a challenge to extract a high enough price to pay for the development costs. If there is no way to exclude others from reproducing and selling an idea, competition will drive the price down to the marginal cost but the average cost of production, which is the sum of the fixed and marginal cost at each point in time, will always be higher than the marginal cost. As a result, it doesn't matter what quantity of the product is demanded, the producer of an idea will never be able to recoup its development costs. So here we reach the crux of the economic or market problem of knowledge-based growth. Unless there is a way to pay for the development of new knowledge, it will not be forthcoming from the market. This creates a significant problem.

In response, governments and industry have developed a number of strategies and institutional approaches to assist innovation. In some cases government acts as a supra-firm, directly undertaking the research and development work in universities and public research laboratories, or by financing the work by private actors through a wide array of direct subsidies and tax credits. A recent study offers an international comparison of income tax support for R&D in Organisation for Economic Co-operation and Development (OECD) countries. The R&D tax systems were ranked by comparing the minimum benefit–cost ratio at which an R&D investment becomes profitable, given a country's income tax treatment for corporations performing this work. That study showed that, after taking account of both federal and provincial incentives, Canada's income tax treatment for R&D investments was the most favorable among the G-7 countries, with an implicit subsidy of about 28 per cent compared to second place Australia, with a 10 per cent implicit subsidy. The US, France and the UK were the only other OECD countries with positive subsidies. Italy, Japan and Germany were the least favorable.[37]

Governments around the world have also worked to develop a comprehensive intellectual property system to make new inventions and creations more excludable. National systems of patent, copyright, trade secrets, plant breeders' rights and trademarks are now being knitted together by international treaty (mainly the 23 World Intellectual Property Organization treaties) and the functioning of the Trade-Related Aspects of Intellectual Property Rights (TRIPS) Agreement of the World Trade Organization. All of these systems grant some form of exclusivity for defined periods, in exchange for some levels of declaration or disclosure. This evolving system of IP protection is bolstered by a variety of special measures in contract and tort law, competition policies and regulatory systems that help to define and limit the uncertainty and liability of firms with new products.

Although part of the response to the knowledge challenge is in the purview of the state, private companies have adopted a number of approaches to increase the benefits from their intellectual property. Three interconnected corporate strategies, which at first blush may appear counter-intuitive, have proven to increase the value of some knowledge-based innovations.[38]

First, firms try to use the 'lock-in' strategy whenever technologies require large investments by customers to develop the know-how to use the technology. In essence, as consumers and firms invest in learning to use a technology, they make sunk-cost investments. If those costs are high enough, consumers may be locked in to using that system even if there are continuing costs, because it would be too costly to switch systems. The ultimate example of lock-in is the layout of the average computer keyboard.[39] The layout of the keys was designed more than a century ago to slow down typists to stop them from jamming keys on a manual typewriter. Few manual typewriters are still used, yet we continue to use the old layout because the cost of each typist retraining on a new keyboard far exceeds any benefits of faster typing. More recently, many computer programs have first been given away for free (at the marginal cost) and then, once demand is established, subsequent versions have been priced successively higher. Most knowledge-based firms would use lock-in if it was possible. In the agricultural biotechnology world, for example, many firms are locked in to using Monsanto's 35S CAMV growth promoter because it is freely available for research, but licenses for commercialization cost more.

Second, firms use aggressive branding strategies to increase the exclusivity of their offerings. Formal industry standards and sometimes regulations that impose equivalence measures also can create the potential for individual firms or groups of firms to control entry into a whole category of products. If a company's property, good or service becomes the benchmark for regulatory, HACCP or ISO standards, that could enable the company to meet the standards at all times but creates a challenge for others. Monsanto's 35S CAMV growth promoter, for example, has become the de facto standard for growth promoters in the regulatory system for GM plants as all crops released to date have used this technology. Regulators are increasingly comfortable with it and do not subject it to the same scrutiny as they do more novel promoters. Moreover, co-branding – two products marketed together, such as Intel chips and Microsoft operating systems in computers – can make it harder for one partner to switch or for new market entrants.

Third, firms are always on the lookout for network effects. Essentially, network effects are those positive externalities that occur when many people use a common technology; although the product is individually bought, a significant external benefit flows to other uses, which increases everyone's willingness to buy the product. Network effects are over and above the general returns to use. For example, the first user of a fax machine, the Internet or a G3 cell phone got limited value because there was no one else to communicate with or content to look at. Once millions were using those services, the value of the technology was much greater. There are ample examples of this phenomenon, especially related to transformative technologies. In the biotechnology world, the rapid uptake and general use of process technologies like *Agrobacterium tumefaciens* mediated transformation, the 35S CaMV growth promoter and genome maps of different species have all generated significant network benefits that are shared between all users and the owners of the technology. In the ICT world, the advent of the telephone, computers, facsimiles, computer software and the Internet have all generated significant network benefits.

The recent focus on knowledge-based growth has highlighted the challenges markets must contend with when confronted with technological change. Although the market is far from powerless in responding, many of the responses of market actors, perhaps consistent with the microeconomic foundations of markets, are seldom made in the impersonal, arm's-length fashion proscribed by the theory.

## THE LIMITS OF MARKETS

Markets are fundamentally about voluntary, self-interested exchange. Nobel Prize winning economist James Buchanan laments the current focus on scarcity that flows directly from Lionel Robbins' classical definition of economics as 'the Science which studies human behavior as a relationship between ends and scarce means which have alternative uses'.[40] Buchanan argues that 'the modern failure is that twentieth-century economists had been converted to the maximization-scarcity-allocation-efficiency paradigm . . . which draws attention away from individual behavior in exchange contract'.[41] He argues instead for economics to focus on the 'origins, properties, and institutions of exchange'.[42]

Microeconomic theory is particularly limited when confronted with significant economic change. Markets are fundamentally challenged by innovation. Douglass North concludes that 'neo-classical theory is simply an inappropriate tool to analyze and prescribe policies that induce development. It is concerned with the operation of markets, not with how markets develop'.[43] Joel Mokyr explains why: 'there usually is, at some level, an [sic] non-market institution that has to approve, license, or provide some other imprimatur without which firms cannot change their production method. The market test by itself is not always enough. In the past, it almost never was.'[44] He notes that increasingly:

> The adoption of a wholly new technology is often the target of long debates and public discourse, unlike many other technical and economic choices. The role of *persuasion* and rhetoric in these decisions is something economists have paid scant attention to, and hence they have not had much success in understanding why, for example, some economies have adopted nuclear power or why some have allowed experimental drugs to be sold and others did not. Furthermore, not all resistance is purely social. There are instances in which the technological 'system' resists a novel and improved component because it does not fit the operation of the whole.[45]

Chapter 8 examines how this debate about new technology has evolved and developed over recent decades.

## NOTES

1. Seabright (2004: 21).
2. The Committee for Study of Innovation (2004).
3. Benacek (2005).
4. Seabright (2004: 15).
5. Arndt (1988).
6. Shafaeddin (2004).
7. Gilpin (1991).
8. Bentham (1780).
9. Jevons (1871).
10. Mill (1848).
11. Walras (1874).
12. Marshall (1890).
13. Walras (1874).
14. Gilpin (2001).
15. Seabright (2004: 21).
16. Coase (1960).
17. Williamson (1981).
18. Dahlman (1979).
19. Williamson (1979).
20. Williamson (1983) defines asset specificity as the degree to which a productive asset has diminished value in second-best uses. The lower the alternate value, the higher is the degree of asset specificity.
21. Joskow (1987).

22. Coase (1960).
23. North (1991).
24. Coase (1937).
25. *The Economist* (2004).
26. Rosenberg (1994).
27. For example, Smith (1776); Mills (1842); Marshall (1890).
28. Taylor (1911).
29. www.valuebasedmanagement.net
30. Porter (1980, 1985, 1990).
31. Cited in Rugman and Verbeke (2002: 769).
32. Porter (1985).
33. Barney (1991).
34. Rugman and Verbeke (2002: 770).
35. Grant (2002: 136).
36. Ibid.
37. Warda (1997).
38. Shapiro and Varian (1999).
39. David (1985).
40. Robbins (1932).
41. Buchanan (1989: 21).
42. Ibid., 14.
43. North (1993). North (1991: 103) offers a detailed description of the Suq markets in North Africa and the Middle East that sounds eerily similar to the market for new knowledge-based goods: 'There are no institutions devoted to assembling and distributing market information; that is, no price quotations, production reports, employment agencies, consumer guides, and so on. Systems of weights and measures are intricate and incompletely standardized. Exchange skills are very elaborately developed, and are the primary determinant of who prospers in the bazaar and who does not. Haggling over terms with respect to any aspect or condition of exchange is pervasive, strenuous, and unremitting. Buying and selling are virtually undifferentiated, essentially a single activity; trading involves a continual search for specific partners, not the mere offers of goods to the general public. Regulation of disputes involves testimony by reliable witnesses to factual matters, not the weighing of competing, juridical principles. Governmental controls over marketplace activity are marginal, decentralized and mostly rhetorical.'
44. Mokyr (1997: 1).
45. Ibid., 2.

# 8. Governing with civil authorities

> The United Nations once dealt only with Governments. By now we know that peace and prosperity cannot be achieved without partnerships involving Governments, international organizations, the business community and civil society. In today's world, we depend on each other.
>
> **Kofi Annan, Secretary General, United Nations**

## INTRODUCTION

Most conversations about and analyses of new technologies begin by focusing on either the role of the state or the operation of one or more of the market actors. This two-dimensional space may be a good place to start, but it is far too narrow to understand and appreciate the complex forces involved in technological and institutional innovation. In our personal lives the state and market do play major roles in how we define our existence, our accomplishments and our hopes, but there is much more that defines us as individuals. Our sense of meaning and purpose is seldom simply defined by the state or market. Even the most authoritarian, idealistic government seldom commands complete subjugation of the individual to the whole. Furthermore, the most aggressive, materialistic person usually has interests that transcend the market. Impersonal state-based policies and structures and anonymous arm's-length market transactions offer little sense of meaning to most of us. Instead, we look beyond the state and the market for context, meaning and a sense of belonging and purpose.

We live in a complex, three-dimensional world. In the world, we engage individually or collectively in a wide range of associations that help to define us. These organizations are defined in a number of ways, but all in one way or another exert authority in the civic domain. They are variously labeled the Third Sector, not-for-profit or voluntary organizations, interest or pressure groups, civil society organizations (CSOs) or non-governmental organizations (NGOs). However these civil authorities are named, they span our lives, from the cradle to the grave, and sometimes beyond.

The oldest and arguably the most important association we all have is with family. Family structures have a series of hierarchies, with different individuals or generations dominating and making decisions in different circumstances. Traditionally, most of the power in family systems was centralized, with the oldest or the strongest dominating and making the critical decisions. Nevertheless, at times others had authority, such as in times of war, during migrations, during hunts or in domestic settings. The concept of family has changed dramatically in recent generations. Although we can still find vestiges of the close-knit extended family clans that governed much of our prehistory (in isolated Inuit communities in northern Canada or Indian communities in the Brazilian rainforest), increasingly many people more narrowly define the nuclear family in terms of closely related bloodlines (siblings, spouses, parents, grandparents, uncles, aunts and cousins). There is less deference to age or physical might and more to social status and the ability to generate income and provide for the family. This evolving family structure has become more narrow and delimited (some have called it the atomic family) but it remains an important social institution. In the US, for instance, about 90 per cent of the population lives in multimember family units and most of the rest, perhaps living singly or in non-family situations, are often closely connected to larger family structures. Given that more than 70 per cent of our measured economic activity is consumed by households, how these families make decisions is important. Although economists model consumption as an economic event, anyone living in a family will recognize that decisions to buy new things are seldom made solely based on a rational, economic basis. This is where any new technology or innovation will ultimately succeed or fail.

Beyond the family, the most important and pervasive association is religion. Estimates suggest that as much as 84 per cent of the world's people are members or adherents in one of over 4300 faith groups, including 2.1 billion Christians, 1.3 billion Muslims, 900 million Hindus and 375 million Buddhists.[1] While faith is fundamentally individual, organized religions since the beginning of recorded history have established and enforced doctrine and exerted their authority to define and direct our temporal lives. In pre-Reformation times, churches defined many of the earliest rules governing both civil relationships and the market. The medieval churches, for instance, controlled most of the critical commercial channels – such as mills, markets, early banking, contracts and travel – and carefully used membership and adherence to the faith to control access to those services. The church used its control of baptism, communion, marriage and burial to define who was able to hold property, engage in commerce and bequeath or pass along their estates. It also levied mandatory taxes (called the tithe) and commanded labor for public works and defense or military campaigns. In that sense, religion often exerted many of the powers of the contemporary state. With the Reformation, the dissolution of feudal structures and the inexorable separation of church and state in most countries, most temporal structures have become the sole responsibility of the state. Most churches now focus almost exclusively on managing doctrines of faith. Nevertheless, religion plays a critical role in many contemporary debates. Doctrine can have a major influence on our economic and political choices. Many of the new technologies being developed and their uses (intended or not) pose challenging questions about the nature of humanity. Agricultural biotechnology has engaged many churches in a debate about our role as stewards of the environment, stem cell research has become inextricably linked to questions of conception and the boundaries of life and the Internet (which generates some of its revenue from pornography and gaming) raises more traditional concerns about sin and virtue.

Beyond the two almost universal authorities of family and religion, people in modern societies belong to a wide range of purpose-built civic structures that go a long way to completing us as social beings. In local communities, education and schooling systems provide a major integrating institution. Although lone scholarship (through home schooling or correspondence) is still possible, much of education is socially defined. Furthermore, apart from artisans, most of our training beyond senior matriculation now involves significant socialization, either through craft guilds, apprenticeships, professional articling, mentorship or formal technical and university degree programs. The bonds and ties we make in these educational efforts often form the foundation for our personal, social and professional lives. Many relationships continue into our working lives, where we interact and mutually support our fellow alumni and peers in our respective careers. Meanwhile, in many societies the extracurricular activities of our children during their school years largely define our leisure time. Coaching or supporting our children's competitive teams, music programs, social programs and artistic ventures bring people together in a wide range of new configurations. Similarly, many citizens engage in local and regional community associations, on local boards and commissions and as members in groups dedicated to sustaining our social, cultural, education, health and environmental infrastructure. The scale and scope of this effort has only recently been acknowledged and counted. A 1989 Gallup poll reported that approximately half of all Americans 14 years or older volunteered some time in that year, generating an estimated $170 billion in unrecorded labor. As far back as 1982 non-profit organizations in the United States employed 8 per cent of all workers and 14 per cent of all service workers, and the sector was the fastest growing over the previous decade. The Comparative Nonprofit Sector Project at Johns Hopkins University studied 37 nations and found not-for-profit entities expended $1.6 trillion in 2002, which if a country would rank as the fifth largest economy in the world.

Many of the associations we engage in also relate to our commercial interests. Most countries have active consumers' associations and many have a variety of grass roots movements to offer commercial services to their members, ranging from retail cooperatives and credit unions to mutual societies for various health and social services. Similarly, most people are at least formally if not actively engaged in associations related to their working lives. Unions command membership in almost every country in the world. The rate of unionization for non-agricultural workers varies widely. The US has about 12 per cent unionization, Japan, 18 per cent, Canada, 30 per cent and some European countries have up to 70 per cent of their workers in unions. Some lesser developed countries have virtually no unions and a few transition economies have virtually all of their non-farm (and even some of their farm) workers in unions.[2] Beyond unions, an increasingly large number of workers belong to professional or peer associations, including most academics, lawyers, accountants, engineers, doctors and architects. All of these groups work in a variety of ways to define our competence and capacity and bound our working lives.

Beyond our individual memberships, the firms many of us work with also belong to collective groups. Many firms are members of chambers of commerce, local business organizations and relevant industrial associations. Firms also increasingly engage in and support various market-making activities, such as industry and product standards definition and enforcement. The array of these engagements is significant and important for innovation and will be examined further in later chapters.

All of these civil authorities are a critical part of our society. They are the root of many of the norms that define how we govern the many vagaries of life. Robert Putnam, in a book entitled *Bowling Alone: the Collapse and Revival of American Community* drew a strong parallel between the gradual decline in the array and scale of many integrative civil structures such as community service clubs and bowling leagues and the increasingly difficult challenges of sustaining a vibrant and effective democracy in the United States. Putnam sought to measure the scale and nature of the stock of bonding and bridging social capital in the US, and looked for influences that might affect the trends. He definitely saw the introduction of modern telecommunications and transportation technologies (including the telephone, automobile, radio, VCR and DVDs) as critical factors. He concluded that the decline in social capital has been a contributing factor in the degradation of democratic debate and the decline of the role of moderates, leading to US politics becoming 'shrill and less balanced'; one need only examine the debates about stem cell research to see the impact of this on new technologies. Perhaps equally important for innovation, he argues that government programming is more effective when people feel engaged and empowered, with lower incidence of opportunistic behavior (such as tax evasion), less intolerance of diversity and more support for liberty.

Although there may have been a relative decline in citizen involvement in many local organizations, there has been an offsetting increase in the number and array of regional, national, international and global organizations that offer new opportunities for people to become involved. Many of those organizations transcend both geography and cultures. The evolution of our society, and the technologies and institutions in it, are increasingly defined and governed by the emerging blend of older, locally based organizations that help to define and nurture local social capital and the newer broadly based and often impersonal entities that bridge often remote groups and find new compromises.

## THE NATURE OF CIVIL AUTHORITIES

Civil authorities, by their very nature, produce key elements that make our economy and society work. The collective or civil sector relies on informal kinships to effect 'satisficing barter' (or what might be called 'reasonable' reciprocity).[3] The instruments of power are far less structured than for the state (coercive administrative institutions) or the market (money, capital and contracts) – the collective relies on culture, moral codes, natural law and social cohesion to effect consensual, often ethically based exchanges and action. Both at the small group size and in the integrated whole, civil authorities either with forethought or serendipitously develop norms and conventions that both create social space but also provide the foundation for governing both intra-organization activities and external efforts. All of these organizations can be viewed as interest groups exhibiting shared attitudes and making claims upon other groups in the society for the establishment, maintenance, or enhancement of forms of behavior that conform to their shared attitudes.[4] These groups interact, develop habits of response, and generate demands or claims on other groups in society. All authorities undertake a variety of similar activities. To succeed, each will inevitably need to identify some unifying assumptions, to develop a common language, to choose an acknowledged set of methodologies, epistemologies or practices, to identify and mobilize a set of members, partners and targets and, ultimately, to develop a set of governing rules that will manage the relationships in the organization.

Civil authorities have been categorized and analyzed in a number of ways. Given that most of the organizations are purpose built, one common way to examine them is based on their primary goals, objectives or purposes. Another way is to categorize them based on their fundamental organizational structure and membership. A third way is to examine them in terms of their outcomes. Each deserves some consideration.

Given the wide range of civil authorities, there has been significant effort devoted by others to categorizing and identifying the structure and function of different types of organizations. At one level, one might distinguish between three foundational types of civil authorities, depending on their underlying objectives – some are strongly values-driven, a second group is fundamentally interests-based and a third class is primarily focused on enforcing epistemological norms.

Those authorities that are strongly values-driven have attracted much consideration. Values ultimately reflect what people want out of life, and what they are willing to stand for. They may be defined in the context of personal attributes (such as honor, integrity, honesty and spirituality), cultural traits (such as ethnic, linguistic or national roots), social aspects (including equality or altruism) or workplace related qualities (often loyalty, contentiousness or professionalism). Generally values are not easily mutable – they are grounded in us by our upbringing, our communities and our experiences. The conventional view is that collective action is and (one might garner from the literature) should be all about reflecting values, beliefs and belonging. This approach can be directly linked to psychologist Abraham Maslow's theory of human motivation.[5] He postulated that as humans meet their basic requirements, they seek to satisfy successively higher needs that occupy a set hierarchy. The four lower, deficiency needs, include physiological needs (food, drink, air, sleep, clothing and shelter), safety or security needs (freedom from violence), love or belonging needs (relationships and companionship in families or communities) and esteem needs (recognition by others and self-respect). His top level, termed a 'being' need, involves self-actualization, or 'what a man can be, he must be'. Although others might challenge the strict hierarchical structure and the ordinal ranking of the five needs,[6] many of the elements make intuitive sense and have been accommodated in other articulations of human motivation. Generally, civil authorities organize themselves to meet the higher-order needs, especially self-actualization, by providing a means of engaging and projecting one's concerns.

In the context of innovation, collective organizations play a role in articulating the views of groups in society. Individuals as citizens and consumers would appear to be uncertain, confused or angry about many of the options being presented by the state or market. In absence of clear signals from citizens or consumers, states and markets are challenged to find appropriate ways to govern new technologies. The literature describes this gap as a need for 'voice.' Collective organizations have moved to fill the gap – especially not-for-profit social action groups that represent segments of society that cannot be adequately addressed in the homogenizing processes offered by the state and market. Though collective organizations are often very effective in mobilizing voice, they frequently complicate the governing process. Each of these organizations often reflects a narrow range of values – the more homogeneous their membership, the more effectively they can define their message.

Challenges arise when competing civil authorities articulate values that are incompatible or difficult to pursue jointly. In those situations, some, if not all, of the civil authorities are bound to be disappointed and frustrated at any compromise, because the nature of their claim is not subject to compromise.

A second group of civil authorities is driven by the interests of their members. Although these institutions may articulate some common values, the fundamental driving force for most of these organizations is to derive some type of benefit for their membership. These interests may range widely, often involving structuring rules to deliver desired outcomes or to redistribute impacts of specific market or public policy outcomes. Cost–benefit analysis often can help to identify the potential impacts of diverging interests and can estimate the scale and scope of trade-offs or side payments that may increase the acceptance of a policy outcome. Much of the literature about this type of civil society focuses on zero-sum conflicts (involving the optimal rate of renewable resource harvesting or non-renewable resource extraction) but the innovation process often involves some very important and often overlooked collective processes of market making. Many not-for-profit associations sponsored by industry are designed to develop and deliver a range of pre-commercial and non-competitive research services (such as development of new general purpose technologies to solve common problems or owning and operating a key piece of technology as a toll good), promote or manage standards setting (including establishing interconnectivity in network technologies such as telecommunications and the Internet or defining standards for product quality and safety) and undertake and supervise general market development for general product lines (such as promoting the efficacy and safety of an entire category of products or technologies). In many cases the underlying benefit or cost can be reduced to monetary terms, which allows for a wider variety of outcomes, as side payments or compensation become possible. Furthermore, many of these authorities have an evolving set of interests that will be played out in a succession of repeated processes. The absence of any fundamental values at stake and the nature of the repeated games these groups are engaged in tends to dampen the rhetorical excesses and lead to more cooperative engagement.

Third, there are many authorities engaged in developing, sustaining or protecting some epistemology or process. The best example of an epistemic community is a professional association (such as for physicians or lawyers) or an academic society (such as the American Association of Agricultural Economists). These epistemic communities have recognized expertise and competence in a particular domain and an authoritative claim to specific knowledge within that domain or issue area.[7] These communities and their members have a shared set of normative and principled beliefs, shared causal beliefs, shared notions of weighing and validating knowledge in the domain, a shared technical vocabulary and a common set of practices associated with a specific set of problems. Many epistemic communities span international boundaries. Unlike values-based social movements and interest-based groups (which are unlikely to share causal beliefs), epistemic communities such as disciplines and professions share causal beliefs and a consensual knowledge base, but do not share principled beliefs or interests (which are the unifying feature of social movements and interest groups).[8] For the same reason, epistemic communities do not cohabitate easily with markets or the state – their foundational commitment to a set of causal beliefs and consensual knowledge is inconsistent with the practical politics of legislators and bureaucratic agencies or the self-interested nature of market actors. Nevertheless, epistemic communities are vital actors in the development and gestation of transformative innovations. Governments, and increasingly industry, frequently turn to these communities in the face of uncertainty. Two subsets of the epistemic community are particularly important in the innovation process. Communities of practice – small, tightly knit groups of people working together to develop and spread knowledge – are often instrumental in taking new scientific and technical discoveries and reducing them to practice. Similarly, networks of practice – professional or craft associations that extend across employers and sometimes sectors – link people who do similar jobs for different firms and help to disseminate tacit know-how to the larger economy.[9]

A second way of examining civil authorities is to examine their organizational structure and membership. As one might expect with self-organizing, purpose-built institutions, there is a wide range of structures. One way to differentiate the institutions is by examining their degree of formality in the system and the openness and fluidity of the membership. Although there would be a spectrum of possibilities for these two factors, one way to start is to postulate extremes of structure: formal hierarchies versus ad hoc groupings; and closed, formal memberships versus open, fluid participation. The resulting 2×2 typology offers four general categories of civil authorities.[10] Primordial networks are those systems with informally structured and limited or closed membership, as characterized by multifunctional craft- or project-based communities, such as in film production, construction, the diamond trade or ethnic communities. The second category, involving closed membership but formal, contractual arrangements is often found in supply chain networks, such as developed and operated by the Toyota Motor Corporation and now emulated in a number of industries around the world. Open membership systems exhibit a different type of structure. Invisible colleges, such as most epistemic communities, involve fluid membership and informal arrangements. These systems depend on fast and universal access to knowledge flows and often are founded on or generate purpose-built research networks or collaborations. Finally, strategic networks allow extensive fluidity in membership but impose restrictive contractual arrangements on those in the system. These types of systems tend to involve purposive relationships (such as pursued by many interest groups) that emerge through active relationships. Many of these systems operate in areas of extensive transformative innovation, such as semiconductors, biotechnology and chemicals. Although some of these systems may exhibit traditional command and control pyramided or hierarchical structures, most exhibit some network features. The key to making these networks work is that they share the same communication codes, allowing them to exchange information and knowledge freely across the networks.[11] The nodes in these highly decentralized systems operate with a large degree of autonomy – the networks are 'flat', with leaders empowered to act under a minimum of direction and supervision – enabling them to be relatively nimble and flexible.[12]

A third way to examine civil authorities is in terms of their outcomes. Putnam argues that the most important outcome of civil authorities is that they generate two important types of social capital.[13] He differentiates between bridging social capital, such as mass movements and broad-based service clubs (and one might add epistemic communities) that are outward looking and encompass people from a wide range of classes and communities, and bonding social capital, such as fraternal organizations and exclusive clubs that look inward and reinforce identities. He likens bridging social capital to a sociological 'WD-40' and bonding social capital to a sociological 'superglue'. Both are important for governing transformative innovation. Bridging social capital (frequently generated in the civil authorities) is vital to generating a language, epistemology and structure for understanding and examining potential transformative innovations, and bonding capital (frequently generated inside firms, government departments and agencies or close-knit purposive communities, such as in research teams) is critical to focusing energies on developing, delivering and managing specific, applied innovations.

## THEORY OF GROUPS

While there has been extensive effort to categorize and analyze the nature, scope, structure and effects of collective action, there has been less effort to find a causal theory of group action. Historically much of the theorizing about groups was normative, assuming the ends justified and generated the means, but now the literature is increasingly looking for behavioral, causal explanations of why some groups form and achieve their goals but in other situations groups fail to emerge.

At one level, we can find some theoretical support for collective activity in anthropological studies. Anthropologists have estimated that our capacity for social interaction with other members of our species has more than doubled from the capacity of our *homo habilis* ancestors but that the optimal size of a modern human's social group is still only about 150 people.[14] This may at least partly explain why in a globalizing world where we are increasingly faced with issues engaging millions if not billions of others, we seek out groups to join. It also perhaps explains why many of the mass movement organizations keep their face-to-face operations small and local and only aggregate their efforts in careful and staged ways. Although some NGOs count their membership in millions, most of those organizations have only minimal engagement by their average member. Those that more fully engage their membership often operate in a distributed or federalist structure, with relatively small, local chapters combining through some hierarchical system into a larger whole. This structure has proven effective for many organizations as they attempt to create a social space.

Beyond the predisposition to seek out smaller groups that fit our capacity, there are three different approaches to how groups form, offered by sociology, political studies and economics.

Sociologists have spent the greatest amount of time and energy examining the underlying forces driving collective action and the role of individuals in those systems. The sociological approach generally starts with the assumption that the individual is at the center of society. Groups are formed and joined by individuals as they seek to pursue their values, interests or livelihoods.[15] The challenge for the study of innovation is that most of the sociological theories have tended to be used to examine other issues, such as cultural identity, gender, ethnicity, crime, family and health. Although these issues may at times be interrelated with innovation processes, for the most part they are not the key issues influencing the scope and scale of any major transformative innovation. The issues examined through the sociological lens tend to be outcomes of the innovation process rather than critical inputs to the system. When they have been considered explicitly in the context of innovation, the approach has tended to be more normative, focusing on how the existing power systems can or should be altered to empower disenfranchised individuals or groups.

Political science differs somewhat in that it has tended to look less at the role of the individual and more at the role of collective organizations in the overall political process. There are perhaps two main streams in the literature. The first 'formal' variant posits that our earliest and most primitive organizations were based on kinship and that, as society grows, organizations take on some of the functions that family previously offered. This model is consistent with Maslow's hierarchy of needs, suggesting in an increasingly impersonal, atomistic world, we are driven to groups in search of validation and self-actualization. The causal variant, in contrast, harkens back to Aristotle and the assumption that man is by nature a political animal, with an instinct, tendency or propensity to herd. Many would date the formal study of organizations in the political system to Arthur Bentley, who in 1908 wrote a book that is credited with initiating group theory in modern political studies. He urged study of the political process in terms of groups, interests, and pressures, where politics and policy were nothing more than reflections of group competition in the political process. David Truman in 1951 revived Bentley's group process theory of government in order to examine the rising role of interest groups in the United States and abroad. This approach offered a tool for analysis, a theory to drive systematic behavioral research and a range of testable hypotheses related to the political orientations of groups, the internal politics of groups and the influence of groups on the legislative, executive, judicial and electoral systems. There has been some examination of groups and politics in the innovation process but much of that work has tended to be too narrowly focused on politics to fully explain their role and impact.

The difficulty with both the sociological and political theories is that they assume it is both obvious and rational for individuals to act collectively. None of them explicitly examines the motivations and choices individuals may be faced with. In 1965, economist Mancur Olson produced *The Logic of Collective Action*, which posed a truly behavioral theory of collective action. He began with a positivist, economic assumption that individuals are self-interested actors who will undertake some form of introspective optimization calculus about whether to join a group or not (similar to what is assumed to happen when individuals make consumption decisions). The challenge is that unlike most rival, excludable private goods, collective action exhibits the key attributes of a public good, in that the resulting benefits are likely to be non-rival and non-excludable. Olson's logic was that if groups were truly public goods, then there would be a real problem with free riding. Although there would be clear and unambiguous benefits of collective action, each individual undertaking their own self-optimizing calculus would be likely to judge that they would be best off if the good is created (by others) but they don't have to invest their energies or resources in creating it. Hence, in absence of some way of compelling membership, many groups would fail to form, and consequently we would get a public goods market failure. As long as the good being pursued is non-collective (in other words, rival and excludable, with no positive externalities), there will be adequate incentives for either a single individual to provide the good, or for a group to form and pursue the good jointly. Whenever the goods being produced are non-rival or non-excludable, then no one individual is likely to produce it and it is problematic whether any group will form spontaneously to produce it. Olson offers a taxonomy of different types of groups and their likelihood of success. He posits that the key to success is the size of the latent group. The fewer individuals required that comprise a group, the more likely it is to form and function. Wherever a single individual would realize benefits at least equal to their costs, the good will be delivered. In areas where there are limited potential members, it is usually possible for groups to join, especially if the goods being generated can be differentiated. This is especially true if there is a potential for repeated processes, as repeated cycles work to lower the costs of organizing.[16] The problems arise when a latent group has a large number of members. In those cases, the costs of organization and the individual benefits of membership are such that it is frequently difficult to organize. These latent groups then have a choice. Even if the good they produce is non-excludable, latent groups may find it possible to organize by offering either selective incentives (such as excludable benefits such as liability insurance for doctors) or by exerting some ability to coerce members in order to survive (for example closed union shops, professional licensing, or compulsory membership). In this sense, the collective action actually becomes a by-product of other purposes.

Although Olson's theory offers valuable insights into the nature of many of the groups involved in the innovation system, there is some debate about how universally applicable it might be. He argues that it also works well for many social action groups, in that social status and social acceptance are individual, non-collective goods whereas social sanctions and social rewards are selective incentives. Hence, privileged and intermediate groups with few members are likely to be able to organize and operate without great difficulty. Thus, the theory goes a long way to explaining the emergence of many of the interest groups, epistemic communities and some of the values-based groups. Olson's logic, however, requires rationality in the economic sense, where individuals undertake some form of selfish optimization calculus. This may not, however, hold true for many philanthropic or religious lobbies, or even for many social-action mass movements seeking to further their values and beliefs. Mass movements are often explained in terms of their 'alienation' from society. Hence, psychological investigations may be the best route for understanding how some of these operate.

## COLLECTIVE ACTION IN THE INNOVATION PROCESS

Olson's theory offers a foundation for understanding the diversity of civic authorities. He notes that collective action may conform with the 'principle of fiscal equivalence', where there may be a need for a separate jurisdiction or governing structure for every collective good with a unique catchment area or domain. Given the widespread diffusion of collective action, his theory would appear to be more right than wrong.

Before proceeding to the next section of this book and the examination of the roles of different actors in governing transformative innovations, it is perhaps worthwhile examining the array and practical structure of many of the civil authorities involved in the area. The Global Civil Society Group at the London School of Economics produces the most comprehensive summary of non-governmental organizations. Their latest reports show that in 2003 there were more 59 000 international non-governmental organizations.[17] More than 70 per cent of all of the organizations appear to operate in areas relevant to transformative innovation. Almost half of the organizations operate in the economic development and research fields, providing the underpinnings for developing and commercializing new innovations. About 12 per cent of the organizations focus on law, policy and advocacy and another 14 per cent work in the broad areas of the environment and social development.

Perhaps what is most surprising about these NGOs is that although many of them are designed to mobilize traditionally disenfranchised voices in the political and economic power systems, they are disproportionately the domain of rich, northern countries. Although these organizations are increasingly drawing membership and support from a more diverse membership base, they are still significantly located and directed by men from wealthy countries. The Global Civil Society Group undertook a survey of the main international NGOs in 2003, and reported that 83 per cent of the almost 18 000 organizations reviewed had secretariats located in OECD countries (53 per cent in the EU alone, with 8 per cent in Brussels, 5 per cent in London, 4 per cent in Paris and 1 per cent in Rome). Only about 12 per cent were located in middle-income countries and just over 5 per cent were in low-income countries.[18] Another survey, this time of 2600 leaders of international NGOs, showed that 77 per cent were citizens of OECD countries (55 per cent from the EU), 7 per cent from Asia, 2 per cent from Eastern Europe, 5 per cent from Latin America and 10 per cent from Africa. Furthermore, almost three quarters of the leaders were men.[19] As one might expect with such a selected cohort, these leaders come from a relatively elite group. A survey in 2002 showed that 91 per cent of the leaders were proficient in two or more languages and 40 per cent were proficient in four or more languages.[20]

In the context of the innovation process, there are a number of high-profile civil authorities engaged in governing a number of recent transformative innovations. Most institutions parse relatively easily into the three domains of values-driven, interest-based or epistemic communities and exhibit a range of openness and formality that generates a mix of bridging and bonding social capital.

Much of the focus in recent years has been on the rise of values-driven civil authorities. These institutions have captured the attention of the media, have aggressively engaged in setting and at times changing the public agenda and have begun to collaborate with both industry and governments to embed their values into market and state governing systems. These institutions vary widely in their structure and operations, ranging from open democratic organizations to idiosyncratic vehicles for engaging personalities.

The Sierra Club and Friends of the Earth epitomize the open, mass-based, democratic approach. The Sierra Club, founded in 1892 by John Muir, is the oldest and arguably the most powerful environmental group in the United States. The club, a tax-exempt organization in the US with income in 2004 of US$92 million and more than 60 chapters in the US and a spin-off organization in Canada, is organized along traditional representational democracy lines, with its executive drawn from the membership and key positions or campaigns presented to the membership for ratification and approval. The organization engages in traditional advocacy and education programs and in direct action through its offshoot, the Sierra Club Legal Defense Fund, a nonprofit law firm created in 1971 to serve as a legal arm to the club's grassroots operation (in 1998, its name was changed to the Earth Justice Legal Defense Fund). Friends of the Earth International, founded in 1971 by four organizations from France, Sweden, England and the US, is now a federation of 1.5 million members and supporters in 70 countries. The organization orchestrates campaigns on a wide range of environmental and social issues. The FoEI is perhaps even more decentralized that the Sierra Club (all its member organizations are autonomous) and similarly democratic: every two years there is a week-long general meeting where the policies and activities of the federation are decided and at which all members have an equal say. These two open, semi-formal organizations both generate bonding social capital (through their local chapters and the engagement of members in their targeted campaigns) and linking social capital (through their engagement with fellow members in their federal systems and through international participation in a number of networks, including the Environment Liaison Centre International, the World Conservation Union and the Economic and Social Council of the United Nations).

Greenpeace, the largest environmental organization in the world with about 2.7 million contributing members and an annual €160 million (2004) operation coordinated through offices in 41 countries, is undoubtedly a values-driven organization, but differs somewhat from the Sierra Club and FoEI in that it is somewhat less open and somewhat more formally structured. When it was started by a group of young activists in Vancouver, Canada, in 1971, it was a small, tightly knit direct-action group targeted on ending nuclear testing and commercial whaling. As it expanded, it added more formal structure, becoming structured in 1979 as a centrally coordinated, hierarchical organization incorporated as Greenpeace International. Part of that arrangement was that overall direction of the movement was centralized in head office and the local offices were tasked with raising and forwarding funds to the international organization. Unlike the Sierra Club and FoEI, campaign targets are not democratically decided. Instead, the organization has adopted a business model where each campaign is centrally chosen in order to be self-financing through donations. This model has kept a sharp focus on a few environmental issues, which recently have included climate change, energy, forests, oceans, toxic chemicals and genetic engineering. Although some centrally planned direct action is still used, sometimes involving local activists, much of the work is now focused on a range of advocacy activities. The organization historically created significant bonding capital (through its close-knit direct action teams) but its centralized structure and focused business model has limited its ability to generate significant linking social capital.

Another group of smaller and more personal, values-driven advocacy groups have engaged heavily in the debate about transformative innovation in recent years. The Foundation on Economic Trends, a US-based, not-for-profit, tax-exempt advocacy organization with an annual budget of about US$450K in 2004, is personified in its founder and president, Jeremy Rifkin. The organization raises funds from interested individuals and undertakes litigation, public education and advocacy. The David Suzuki Foundation, an environmental non-profit organization registered in Canada and the United States, is similarly personified by its eponymous founder. The organization, founded in 1990, reports it has 40 000 contributing members, a staff of 35, and a range of projects targeted on climate change and global warming, ecosystem-based forestry, sustainable fisheries, aquaculture and biodiversity. The Foundation communicates its findings through speeches by Dr Suzuki, a series of publications and in advocacy efforts. Both organizations, and a wide range of other focused groups of the same type, are relatively closed, formal systems that are unlikely to generate much linking or bonding social capital.

There is a range of traditional and new interest-based civil authorities engaged in the debate on transformative change. In the area of biotechnology, for instance, the industry is well organized. In the US, the Biotechnology Industry Organization (better known as BIO) was formed in 1993 with the merger of the Industrial Biotechnology Association, which primarily represented larger, established companies and federal regulatory agencies and the Association of Biotechnology Companies, which represented emerging companies and universities. BIO in 2004 reported a membership of 1131 companies from across the spectrum of biotechnology applications, academic institutions and biotechnology centers. The organization offers an array of services, including advocacy with the US regulatory agencies, but is probably best known for its annual convention and exhibition, the largest biotechnology event in the world which drew more than 17 000 participants from more than 60 countries in 2004. This organization generates significant linking social capital but its open and relatively informal structure diffuses its focus and limits its ability to generate much bonding capital. CropLife International, in contrast, is a global federation representing the largest plant science companies in the world, with a worldwide network of regional and national associations in 91 countries. Unlike BIO, CropLife has a relatively small and homogeneous group of members. Although each of the 91 national organizations has a range of local participants, the organization is clearly the advocacy arm for the large crop protection and pest control companies, and commands significant influence in public debates about the regulation and management of the sector. This closely knit organization is thereby able to generate significant bonding capital for the industry.

There is also a wide range of commodity and industry groups interested in the impact of new technologies on their industries. One example, the Canola Council of Canada, initially established in 1968, is a Canadian-based, sectoral-focused, industry-led association that uses funds derived from a variety of sources (including public and private grants and a voluntary industry levy on canola processors and exporters) to fund targeted technology research and development, to provide a venue for knowledge sharing and coordination in the research system, to undertake national and international market development, to provide agronomic advice to farmers and to establish industry standards for canola quality. This organization is often credited with having created very important linking and bonding social capital in the canola industry in Canada, to the point that it is one of the most organized and aggressive sectors in the Canadian agri-food system.

Consumer-led, interest-based groups are also becoming more active in the debate about new technology. Most operate as large, relatively open, democratic national or international federations. In the US, for instance, Consumers Union, publisher of *Consumer Reports* since 1936, is an independent, non-profit testing and information organization serving consumers. The organization issues one of the most comprehensive and unbiased reports offering advice about products and services, personal finance, health, nutrition and other consumer concerns. The Union, governed by a board of directors elected by CU members, also advocates on behalf of consumers' interests. The Canadian equivalent, the Consumers' Association of Canada (CAC), was founded in 1947 as an independent, not-for-profit, volunteer-based, charitable organization with a mandate to inform and educate consumers on marketplace issues, to advocate for consumers with government and industry and to work with government and industry to solve marketplace problems. The CAC focuses its current work in the areas of food, health, trade, standards, financial services and communications. The organization operates within a federal structure, with provincial associations and local member participation. CAC volunteers are active on a wide variety of internal issues committees, consumer/industry/researcher networks, national task forces and governmental processes related to a wide range of technology issues, including standards setting processes. Both organizations, being open and democratic, offer the prospect of creating significant linking capital and some bonding capital among those who engage in the democratic process and in the committee work.

One group of actors that exhibits both values and interests and is becoming increasingly important, particularly in the biomedical area, is the patient advocacy organizations. The American Cancer Society, for example, is heavily involved in funding and governing the research related to cancer detection and treatment. Founded in 1913 as the American Society for the Control of Cancer by 15 prominent physicians and business leaders in New York City and reorganized in 1945 as the American Cancer Society, it is now undoubtedly the pre-eminent nationwide community-based voluntary health organization dedicated to eliminating cancer as a major health problem by promoting research, education, advocacy and service. With more than two million volunteers nationwide, a US$820 million budget in 2004 and more than 3400 local offices, the American Cancer Society offers a number of programs related to cancer prevention, early detection, and patient service at the community level and plays a major role in the cancer research community (where it provides funds, is part of efforts to set and administer research policy and promotes public and firm support for cancer research). Beyond the services it offers, its group of active volunteers has gone a long way to generating a linked community of engaged, interested citizens who have in many cases gone on to generate new and different patient advocacy organizations.

The group that is perhaps least studied, and is not even in the statistics offered by the LSE Global Civil Society Group, is the epistemic community. A wide range of groups fit in this category, including the pinnacle royal or national academies, the disciplinary associations and even the professional and industrial standards setting bodies.

At the top, there are a group of about 80 national or international academies of leading scholars which are often called upon to strike committees to offer opinions on emerging issues. The world's oldest scientific academy in continuous existence is The Royal Society in the UK, which was founded as an independent body by Royal Charter in 1660. The backbone of the Society is its Fellowship of the most eminent scientists of the day, elected by peer review for life. There are currently 1400 Fellows and Foreign Members. The Society's goal is to support 'excellence in science' through grants to young scientists, via advocacy on behalf of education and research in the sciences and mathematics, by providing independent and objective scientific advice on a wide range of issues (through a peer-mediated policy statement process), by convening academic meetings, through the publication of seven peer-reviewed international journals and with an array of medals and prizes.

While the national academies may set the goals and deal with controversies, the estimated 17 500 scientific scholarly societies operating in the world today do the bulk of the work in governing the epistemic communities.[21] These societies manage and publish the estimated 50 000 peer-reviewed journals currently being published.[22] The American Economic Association is one example of a pure epistemic community, with a clear goal of encouraging economic research, sponsoring the publication and dissemination of research findings and encouraging freedom of economic discussion. In 2004, the Association had a membership of approximately 19 000 economists (the majority in academia but a large number of others from industry and government) and had about 4500 libraries, institutions, and firms subscribing to one of the leading quarterly publications of the Association (*American Economic Review*, the *Journal of Economic Literature* and the *Journal of Economic Perspectives*). In addition to the largely non-rival, non-excludable benefits of promoting research and peer-reviewed publication, the association provides a monthly jobs posting, bestows prestigious awards on notable members and holds the pre-eminent annual academic congress in the discipline (which offers a market for both ideas and people). Open to all subscribing members and governed by an executive ratified by universal ballot, this association generates significant linking social capital within the economics profession.

The American Medical Association, founded in 1847, represents a somewhat different epistemic community – it both develops public goods and generates professional standards which act as selective, excludable incentives for membership. As the pre-eminent organization uniting physicians in the United States, the organization undertakes many of the functions of a epistemic community (such as holding academic congresses and publishing journals) and acts as an advocate for health policy. Moreover, through the operation of a number of its councils, it maintains and updates the 160-year-old code of ethics for the profession, is responsible for recommending the appointments of representatives to accrediting bodies and is an accredited sponsor of continuing medical education. This array of activities, similar to those delivered by many other professional associations (including engineers, lawyers and chartered accountants), both directly governs parts of the system by control of professional standards and generates significant bonding social capital among their associates.

A final epistemic group which is almost always overlooked is the array of industrial standards setting bodies. The ISO/IEC Guide 2:2004 defines a standard as a 'document, established by consensus and approved by a recognized body, that provides, for common and repeated use, rules, guidelines or characteristics for activities or their results, aimed at the achievement of the optimum degree of order in a given context'. Each standard covers one or more disciplines, is coherent and consistent, results from participation, is a living process that needs to be kept current, has a reference status in commercial contracts and in court, has national or international recognition and is available to everyone. Although they are unambiguously sustained in the interests of their members, functionally they operate like most epistemic communities, focusing on a set of processes and principles that they define and protect. Although the International Standards Organization (ISO) is widely viewed as the pinnacle standards organization in the world, it is one of more than 50 international standards organizations operating in the world today.[23] Furthermore, each of these organizations has a federal structure, with national and sometime sectoral affiliates. ISO, for example, has 146 national standards bodies as members, each which operates a similar hierarchy of relationships in their own countries. The ISO system alone has more than 15 000 operating standards and more than 2950 technical bodies working on sustaining the existing standards or developing new ones. Standards setting organizations are constitutionally formal and transparent and generally are relatively closed forums where the specific elements of a standard are worked out between those integrally engaged in the production or use of a product or service. In that sense, they have the potential to generate both linking and bonding social capital among those engaged in the extensive standard setting process (although in contentious cases, such as the Canadian Council of Grocery Distributors' process to establish a standard for genetically modified foods, social capital could potentially be destroyed).

## CIVIL AUTHORITIES AND CORPORATE SOCIAL RESPONSIBILITY

Many values-based civil authorities have recently pressed both governments and industry to internalize into their corporate or organizational goals a set of guiding principles and values and to report on activities and performance related to those new objectives. This effort has been variously labeled as 'the third way' in government and 'corporate social responsibility' in industry. Each movement reflects an unease with monopoly state action or unbridled markets. Anthony Giddens, in his 1998 book *The Third Way* that was extensively quoted by Tony Blair and Bill Clinton, argued for a new approach to government that combines 'social solidarity' with a dynamic economy. This new 'dialogic democracy' would involve proactive and reflexive individuals and social groups in dialogue with each other and with government. On the corporate side, the concept of corporate social responsibility (CSR), sometimes called the triple bottom line, has evolved to signify the commitment by companies to pursue, measure and report to stakeholders and society on profits, environmental protection and social justice. Although this concept predates the Third Way movement, they converged in 1997–8 when Social

Accountability International was created. It established its first multi-stakeholder standard and certified its first company (an Avon factory in New York in 1997). These parallel processes have generated what Michael Power from the LSE has characterized as an 'audit society' where auditing is becoming a principle of social organization.[24]

Power argues that audits are changing the nature of our governing systems, as they are increasingly concerned with systems in place to govern quality and not with outcomes themselves. Essentially, what gets measured gets done. He observes that there has been a shift in the style of audit. The traditional model addresses low-trust situations where formal discipline is exercised in the system with quantitative, *ex post*, long-distance, single-measure audits by private, expert, external agencies. In contrast, he sees the Third Way/CSR movement increasing the demand for an alternate audit style, where high-trust, autonomous systems are evaluated using qualitative, real-time control, local, multiple-measure audits done through a combination of public dialogue and internal agencies.

This shift has triggered both gains and losses. In the government sphere, regulatory and fiscal authorities tend to be clearly delineated in law in specific administrative or technical terms, which has tended to create a disconnect between social accounting and formal audits offered by government auditors general. Meanwhile companies are being challenged to develop meaningful measures for environmental sustainability and social justice. The OECD has even developed a set of guidelines for multinational enterprises in an effort to spur 'responsible corporate behavior worldwide', laying out a set of concepts, principles and guidelines related to disclosure, employment and industrial relations, the environment, bribery, consumer interests, science and technology, competition and taxation.[25] Although many of these suggestions relate to general good business practices, the guidelines offer an aggressive set of objectives for firms. A ranking of companies using such measures, produced by the consultancy SustainAbility, the UN Environment Program and Standard and Poor's, suggested firms were using a mix of different methods to reflect their efforts. So far there is no convergence. A survey by the Economist Intelligence Unit and Deloitte asked 249 executives in 2004 about their use of social reporting. Almost three quarters said the companies are under pressure to use non-financial reporting measures, but they almost all agreed that the available measures are 'mediocre or poor'. Only 9 per cent of respondents saw these measures as appropriately valuing the firm's efforts.[26]

## CONCLUSIONS

A few points about the role of civil authorities in the broader social debate are worth highlighting. In the first instance, it is important to understand the foundational driver for the groups. Although interests can be traded or compensated, values and epistemologies are less likely to be malleable or negotiable. This basic fact offers some insight into the difficulties of dealing with civil society in the governing process. Keeping in mind the IAP2 public participation spectrum introduced in Chapter 6, it is clear that engaging diverging values-driven or epistemic civil groups in higher-order participation can lead to clashes, as values conflict or epistemologies are challenged.

Second, in the context of transformative innovations, all three types of civil authorities may be valuable actors. The epistemic communities are clearly fundamental to understanding the underpinnings and nature of a transformation, its position in our knowledge system and its potential environmental, public health, economic and social impact. Similarly, many interest-based authorities are the key drivers for creating many of the conditions for a market, including engaging in pre-commercial, non-competitive research to overcome technology hurdles, defining technology and product standards to facilitate interconnectivity and adaptability in supply and demand systems, and undertaking the vital investments to test, evaluate and adapt new technologies and products to our economy and to our society. Nevertheless, there remains a vital role for values-driven groups to assist policy makers, firms and individuals (as consumers, citizens and members of communities) to become aware of the opportunities and challenges, to grasp the scope of options and to mobilize and articulate collective, values-based views about the appropriateness and suitability of transformative innovations.

The rest of this book will examine how the multitude of civil authorities has engaged with states and markets in an effort to govern recent transformative innovations.

## NOTES

1. www.adherents.com
2. ILO, World Labour Report (1997–8).
3. Benacek (2005).
4. Truman (1951: 33).
5. Maslow (1943).
7. For example, Wahba and Bridwell (1976).
8. Haas (1992) introduced the concept; Mokyr (2002a) offers a good discussion of their operation and structure.
9. Ibid.
10. Brown and Duiguid (2000).
11. Powel and Grodal (2005: 64).
12. Castells (1996).
13. Stein (2001).
14. Putnam (2000).
15. Homer-Dixon (2001). In contrast, Olson (1965: 54) argues that even 150 people may be too

large for many purposes. He suggests that very small groups are the most likely to act rather than to debate or seek points of view or reaction. He cites (54) studies from government and industry that show most 'action' groups involve fewer than seven individuals. Larger assemblies (e.g. large committees or parliaments) tend to eschew action for debate.

16. Anthony Giddens, one of the leading sociologists of current times, identifies four theoretical strands in the discipline. Each has a particular focus on understanding structure and order in society. Functionalism looks at how human-made institutions (for example schools, churches, firms) operate individually and interact or combine to promote social stability. In that sense, it is one way to break down social systems into their component parts to examine their role in contributing to distributed governing systems. What is less clear, however, is how the parts (be they formal or informal groupings), form and operate. Structuralism, in contrast, is drawn from philosophy and the study of linguistics, and looks at the role of linguistic, social or cultural structures and their role in creating meaning. This approach has tended to have limited use in the study of political and economic systems – its general focus on kinship systems has at times been used to examine the small group structures that permeate innovation systems. Symbolic interaction delves deeper into the role of individuals, focusing almost exclusively on face-to-face interactions and their role in creating groups, which give meaning and order. Finally, Marxism operates at the level of class, examining the role of class in the power system and the influence of class divisions on conflict. A recent post-modernist movement asserts that constant change is now the status quo, with the result that structures are unlikely to have enduring meaning. This has led to efforts to reassert the role of the individual in the study of society.
17. Ruttan (2003: 204).
18. Compiled from Record 18, Kaldor *et al.* (2004).
19. Author's calculations using data from Record 14, Kaldor *et al.* (2004).
20. Author's calculations using data from Record 20, Kaldor *et al.* (2004).
21. Kaldor *et al.* (2004: 356).
22. www.scholarly-societies.org
23. www.libraryjournal.com/article/ca374956.html
24. World Standards Services Network, www.wssn.net
25. Power (1994).
26. OECD (2000).
27. *The Economist* (2004), 'Does it add value' 13 November, p. 81.

# PART IV

# Governing in practice

# 9. Governing knowledge

Knowledge is power

**Francis Bacon**

## INTRODUCTION

Isaac Newton wrote: 'If I have seen farther it is by standing on the shoulders of giants.' In his day, finding the giants on whose shoulders to stand must have been somewhat easier, as there were only a few sites of higher learning and a relatively small number of thinkers, scholars and inventors. With the advent of the earned doctorate in the nineteenth century and the sharp escalation in university enrollment and public and private research in the past century, there are now literally millions of scholars and practitioners who may be the source of new knowledge and ideas. This profusion of effort makes it quite difficult to catalogue, assess, integrate and disseminate the mass of knowledge that we know or might want to know.

There are many actors involved in governing the normalization of emerging knowledge in the natural sciences, applied technologies and social sciences, making it increasingly difficult to establish what is known, what is new, where it is going and how it might be applied safely and efficaciously. Sheila Jasanoff suggests that knowledge is ultimately 'co-produced' by many constantly intertwined actors at four stages: emergence, contestation, standardization and enculturation.[1] Two main types of epistemic-based knowledge are particularly challenging. Know-why knowledge about the physical and social world around us – largely embodied in academic treatises – and tacit, know-how knowledge – embodied in individual scholars and practitioners – provide the foundation for what we know. As discussed in Chapter 8, both types of knowledge involve codifying and preserving a shared set of normative and principled beliefs, shared causal beliefs, shared notions of weighing and validating knowledge in the domain and a common set of practices associated with a specific set of problems. This process is embodied in both the conceptual framing of 'normal science' and the practice of individual governing bodies in the academies and professions. Transformative technologies, however, frequently involve challenges to the paradigms encompassed in normal science. Obviously, not every new idea or concept will be useful or sustainable. The process of assessing and weighing new knowledge is challenging. In many cases the response to new ideas will be purpose-built processes to address the unique epistemic foundations and institutional structures in the specific knowledge domain.

Over the millennia, society has developed a series of governing systems (some might call them subroutines) to undertake the task of codifying, assessing, integrating and disseminating the knowledge stock. Many fear that the traditional peer-review system of managing the normalization of emerging science has not kept pace with the ICT revolution. In the past new scientific developments were tried, tested and challenged in a slow, methodical way outside the limelight. The goal of science is no longer simply discovery and dissemination of new knowledge; increasingly scientists want broader recognition and commercial gain. More than half of research in many domains is now undertaken either by private sector scientists or by public sector and university scientists funded by, and working collaboratively with, private interests. Now, each new discovery (touting either new unimaginable benefits or unfathomable risks) is put in front of a largely scientifically unsophisticated public with great fanfare through glowing press releases to traditional media outlets and on the Internet, hyped lead articles in learned journals, perfunctory front page stories in tabloids and quality broadsheets and in brief, sensationalized sound bites on radio and television. Tim Caulfield, a legal ethicist at the University of Alberta, has characterized this process in the biotechnology area as 'genohype'.[2] A circus-like atmosphere characterized by excitement and fear has replaced the traditional period of contemplation, assessment and testing, complicating our understanding of new technology.

## CATEGORIES OF SCIENCE

Thomas Kuhn, in *The Structure of Scientific Revolutions*, laid out a useful framework for thinking about the structure of existing and staging of new knowledge. He provides a generalized picture of the process by which new knowledge is born, undergoes change and development and eventually becomes normalized. Scientific development according to Kuhn's model involves the study of phenomena which lead scientists to develop explanatory theories that, after an appropriate period of challenge and testing, lead to a 'mature' science and, if sustained through subsequent challenges, become paradigms. As a matter of course, the evolutionary process involves a period of 'normal science,' when efforts are made to articulate the paradigm, explore possibilities within the paradigm, use the theory to predict facts, solve scientific puzzles, and develop new applications of theory. Throughout this period, science is challenged to address both known phenomena and unknown possibilities. This period of new evidence and expanding application of theory all takes place within the paradigm of normal science.

The state of normal science is challenged when an unknown phenomenon is observed. For the most part, existing theory and methodology usually can adequately and successfully explain these observations. At times, however, there may be further discoveries of natural phenomena that violate the expectations that govern normal science. This involves discovery of problems not previously known to have existed. In some cases, normal science cannot fully explain them. A relatively abrupt shift or transition to a new paradigm would then become necessary, which would require someone to define and exemplify a new conceptual and methodological framework.

Kuhn's examples of the formation and transformation of paradigms were drawn entirely from the history of the physical sciences but the typology can be applied to almost any discipline that seeks to frame itself as science. Both theory and evidence, fundamental features of any science, can be known or unknown. Thus, a classic quadrant box can be used to describe the relationships between known and unknown theory and evidence, highlighting four possible situations of science, which have different membership and control mechanisms that can vary in importance depending on the science involved (Table 9.1)

First, conventional or normal science operates where there is a comprehensive paradigm (with both theory and methodology) that offers predictions that are borne out in the evidence. In these cases, it is possible that new evidence or enhancement of the theories or methodologies could offer finer conclusions, but the general conclusions remain constant. This science is almost exclusively the preserve of the global academic scientific community, having been developed, stored and transmitted from universities and other peer-reviewed research centers. The internal control systems of the global scientific community – namely the peer-reviewed structure of appointment, tenure, promotion and publication – enforce rigor and conformity, reducing the chance for unsubstantiated theory or evidence being perpetuated over long periods. During periods of stability, this science dominates.

*Table 9.1 The four realms of science*

| | Known impact/evidence | Unknown impact/evidence |
|---|---|---|
| Known causes/theory | Conventional science | Hypothetical science |
| Unknown causes/ theory | Experimental science | Speculative science |

*Source*: Modified by author from Kuhn (1970).

As one should expect from Kuhn's description of how science evolves, there are other realms of scientific investigation. Experimental science is perhaps the most frequent challenge to conventional or normal science. As detection systems become refined (involving new genomic, proteomic and metabolomic methodologies), there is an increasing likelihood that new evidence may be found that is not in the first instance explained by the conventional theory. This evidence could either challenge theory directly by the discovery of the presence (or absence) of some element that theory says should have been not found (or found). It could also challenge the theory indirectly by illuminating some impact that is not formally incorporated into the theory. This stage of science has historically been very important, as scientists have sought to discover phenomena that need explaining. The challenge is that while some of this experimental work is done in the academic world, much of it by necessity is done by practitioners or laypersons working in the market and society. These practitioners tend to be isolated from the global scientific community, as they are examining *in-situ* events that may only have local or regional impact. The difficulty in dealing with this science is that the internal governance/control mechanisms are often relatively weak. Measurement and detection methodologies vary widely depending on what they are designed for. The evidence is sometimes but not always fully disclosed and quite often there is no formal peer-review structure to assess the significance of new experimental results, either in statistical or scientific terms. This raises the possibility of erroneous measurement of irrelevant evidence that is widely disseminated through the industry or public press or through less formal channels, such as NGO newsletters and the Internet.

Challenges can also come from the realm of hypothesis-driven science. As new knowledge and techniques accumulate, they raise new questions about the assumptions and structures of normal or conventional science. The challenges range from fundamental to incremental. New hypotheses often emerge when different parts of normal science are combined to examine issues where conventional science may be silent. The challenge is that one can never prove any hypothesis or theory – one can only gather evidence that disproves a hypothesis. Although most influential hypothetical science is generated within the academic world, which ensures that there is a built-in system for evaluating and testing its relevance, two recent trends have complicated that. In the first instance, many fields of science have become part of big science, which has made it more difficult for challenges to surface and be tested solely within the academic scientific community. The price of admission to 'big science' is the ability to pair theory with empirical methodologies, which frequently means working with conventional science. There are few funders – public or private – that will make substantial investments in highly uncertain science. Meanwhile, there has been a rise in the number of scientists with advanced degrees working outside academia (in industry, government and NGOs), which, paired with increased competition by the media and some of the more commercial scientific journals, often leads to premature public release of incomplete or often readily refuted hypotheses. These scientists may have an interest in engaging in the scientific process but are not as bound by the rigors of the scientific method as those inside academia. Hence, we often see the promotion of hypotheses of benefits or risks of new technologies that upon further consideration and experimentation are rejected. Although new hypotheses are an important part of challenging our understandings, they must be treated with care as there are probably an infinite number of possibilities to consider.

Finally, there is the realm of speculative science, where there is neither well articulated theory nor any body of evidence to suggest that a causal relationship is present. This subset of science offers great promise but also poses tremendous challenges. It is probably fair to say that many of the great breakthroughs in science have come from speculation that is not linearly tied to either any specific evidence or to the conventional science of the day. Although this science is important, it most certainly needs to be filtered through further experimentation and theorizing before it can provide a base for action. The difficulty is that speculative science is perhaps the most exciting and engaging of all the realms. Hence, it generates significant public scrutiny and debate often at times when it needs to be considered more fully in the context of the scientific method. The regimes for governing this process are frequently undefined initially; often the structures need to be purpose built to handle the process. Academics engaging in speculative science remain bound by the strictures of peer review but because speculation is often out of the bounds of normal science, it is hard to find a suitable forum for consideration. Equally important, many marginally connected academics and non-academic scientists have nothing to lose by speculating – in the parlance of the day, any media coverage is good, as long as they spell your name right.

One way to examine the systems that govern knowledge is to distinguish between those processes that govern normal, conventional or paradigmic science and those that exist or are built to handle challenges to that scientific base.

## NORMAL SCIENCE

To draw on a sports metaphor, normal scientific knowledge is governed by playing both the man and the ball. Individuals who claim knowledge and expertise are for the most part accredited through an educational or training system (such as a university, technical college, professional body, craft union or guild) while the information, knowledge and ideas generated by these people are similarly vetted through a variety of evaluation processes.

Both processes are founded on a particular framing of the scientific process that is derived from the past four centuries of practice. Historians Stephen Shapin and Simon Schaffer (1985) trace the current structure of science to its roots in the 1660s. They have identified many of the attributes of the modern, scientific experimental method to Robert Boyle and his experimentation on the air pump. They assert that Boyle proposed that matters of fact should be the outcome of a process of engaging in an empirical experiment, which would necessarily include repeated 'witnessing' – 'an experience, even of a rigidly controlled experimental performance, that one man alone witnessed was not adequate to make a matter of fact'.[3] Boyle's focus on the social aspects of science was an effort to remove science from the realm of alchemy, where individuals claimed great accomplishments but could neither replicate nor explain what happened. In his approach, witnessing included two elements: the repeated undertaking of experiments and the experimental essay. The essay, to be written in a plain, ascetic, unadorned (sometimes called functional) style, was designed to provide adequate information to allow replication and to lay out the specific, narrow findings and implications of the experiment (also including experimental failures). Boyle also sought to define the rules for civilized scientific discourse, in an effort to remedy the fissions of the just-ended English civil war. He strongly asserted that disputes should be about findings and not persons. The *ad hominem* style was to be avoided. A corollary of addressing the findings and not the author is that no one was to have a privileged status – the experiments and findings were to be the final arbiter of facts. Shapin and Schaffer note that Thomas Hobbes, a contemporary scientist, defended the earlier humanist, philosophic model of science, asserting that witnessing was not critical, that repetition of known facts was not needed and that the form of debate need not change. In essence, Hobbes envisaged a science with 'masters' but Boyle anticipated a science governed by peers. Shapin and Schaffer argue that both men put forward 'different ways of organizing the production of knowledge', offering radically different solutions to which propositions were to be viewed as meaningful, which problems should be addressed and where to draw the boundaries between known and unknown effects. In the end, Boyle's model has dominated in the physical and natural sciences. Those who declined to play by the new rules were not viewed as legitimately part of science.

This change in the rules of science arguably triggered the scientific revolution. The rules of the game have continued to evolve. The current structure of science and its definition of knowledge is well codified by Michael Polanyi, Robert Merton and Karl Popper, three great philosophers of science.

Michael Polanyi, a scholar of the philosophy of science who drew on his own experiences as a medical doctor, a physical chemist and an economist, argues that science is organized much like a republic: 'the authority of scientific opinion remains essentially mutual; it is established between scientists, not above them. Scientists exercise their authority over each other.'[4] He argues that provided the standards of professional merit are adhered to, the pursuit by independent, self-coordinated individuals will be the most efficient organization of science (this is analogous to the economic framing of society, where self-interested, optimizing individuals are posited to lead to socially desirable outcomes). Polanyi identified three standards of professional merit. First, work must be plausible, such that the structuring or results of an experiment must bear some correlation to established knowledge; outliers, though perhaps correct, will be treated as anomalous. Second, scientific value is assessed by a balance of accuracy, systemic importance and the intrinsic interest of a subject matter. Third, work should exhibit originality. Polyani argues that the balance of the three merits keeps science functioning – while plausibility and value tend to be conservative traits, originality keeps science current and progressing.

Sociologist Robert Merton went further, suggesting that science fundamentally involves: (a) a set of methods by which knowledge is certified; (b) a stock of accumulated knowledge; and (c) a set of cultural values and mores governing scientific activities.[5] He argues that:

> the ethos of science is that effectively toned complex of values and norms which is held to be binding on the man of science. The norms are expressed in the form of prescriptions, proscriptions, preferences, and permissions. They are legitimatized in terms of institutional values. These imperatives, transmitted by precept and example and reinforced by sanctions are in varying degrees internalized by the scientist, thus fashioning his scientific conscience . . . Although the ethos of science has not been codified, it can be inferred from the moral consensus of scientists as expressed in use and wont, in countless writings on the scientific spirit and in moral indignation directed toward contraventions of the ethos.[6]

Merton identified four institutional imperatives driving modern science – communalism, universalism, disinterestedness and organized skepticism – which he abbreviated into CUDOS. Science should be commonly owned (communal) such that 'the substantive findings of science are a product of social collaboration and are assigned to the community'.[7] In a perfect world, 'property rights in science are whittled down to a bare minimum by the operation of the scientific ethic. The scientist's claim to "his" intellectual "property" is limited to that of recognition and esteem . . . secrecy is the antithesis of this norm.'[8] Science is universal in that 'truth claims, whatever their source, are to be subjected to pre-established impersonal criteria: consonant with observation and with previously confirmed knowledge. The acceptance or rejection of claims entering the lists of science is not to depend on the personal or social attributes of their protagonist . . . Objectivity precludes particularism.'[9] In short, good science should be 'international, impersonal, virtually anonymous'.[10] Good science is also disinterested. 'Involving as it does the verifiability of results, scientific research is under the exacting scrutiny of fellow experts. Otherwise put . . . the activities of scientists are subject to rigorous policing, to a degree perhaps unparalleled in any other field of activity.'[11] Finally, science should exhibit organized skepticism as 'both a methodological and an institutional mandate'. In a world where science adheres to the CUDOS norms, the free competition of ideas would trump special interests.

Philosopher Karl Popper honed in on what he judged as the central problem in the philosophy of science – demarcation, or what distinguishes science from what he terms 'non-science'.[12] He draws the line at whether a discipline uses inductive or deductive reasoning. He asserts that science cannot rely on induction, because it is possible to find evidence to support almost any assertion. Rather, falsifiability is the mark of true scientific investigation. Every genuine test of a scientific theory, then, would logically be an attempt to refute or to falsify it and one single, genuine falsification would lead one to reject an entire theory.[13]

Combined, Polanyi, Merton and Popper define the critical elements of the 'scientific method' currently used by most of the positivist physical and social sciences. Although there is no definitive statement of what comprises the scientific method, most disciplines recognize four relatively well defined elements to the method. First, work often begins in an inductive way, with observation and description of a phenomenon. At this stage, the goal is to ask a meaningful question or identify a significant problem, and to state the problem or question in a way that it is conceivably possible to answer it. As one might imagine, this is where normative influences can hold significant sway, as a person's interests and concerns will likely influence the nature of the question. Second, the method requires articulation of a theoretically driven hypothesis to explain the phenomenon. In most disciplines, the hypothesis takes the form of a causal relationship, which can usually be empirically or mathematically defined, tested and refuted. Third, the hypothesis would be used to predict the existence of other phenomena, or to predict quantitatively the results of new observations. Fourth, the method fundamentally depends on performance of experimental tests or empirical examination of data in a structured way in order to validate or refute the predictions. The power of the 'hypothetic-deductive' scientific method is that it provides a process of establishing refutable assertions that can be open to peer evaluation and challenge. If a series of repeated, independent experiments or analysis does not refute the hypothesis, then it may come to be regarded as a theory or law, and thereby enter what Kuhn called a 'scientific paradigm'. If the experiments offer evidence that does not support the hypothesis, then it must be rejected or modified. It is a truism that theories can never be proved, they can only disproved. There is always the possibility that a new observation or a new experiment will conflict with a long-standing theory, forcing a revaluation and restatement of theory.

One final distinction can be made in the general scientific process. On the basis of the epistemological characteristics of knowledge formation, one can distinguish between analytical or science-based knowledge and synthetic or engineering-based.[14] Analytical or science-based propositions tend to be theory driven whereas synthetic knowledge is trial driven. Science-based knowledge can also be characterized as primarily related to exploring and codifying patterns in the behavior of nature whereas synthetic technology refers to transforming our observations of the world to fit some preconceived end.[15] Thus, science or analytic knowledge formation aims to break down systems while synthetic knowledge formation seeks to integrate or construct systems. In practice the two are often intimately interwoven and interdependent but the structures that govern them are qualitatively different.[16]

## INSTITUTIONS GOVERNING NORMAL SCIENCE

Joel Mokyr suggests institutions play a central role in sustaining the organizations in which knowledge is 'expanded, preserved and diffused' by creating and sustaining the 'rules by which they play (such as open science, credit by priority, reproducibility of experiment, rhetorical rules, acceptance of expertise and authority)'.[17] A range of institutions have been engaged in this process, in the first instance defining what we know (and by definition, prescribing what is truly novel) and second creating bridges between different knowledge domains. Five main interlocking and overlapping structures have evolved to handle the task. At the most basic level, groups of scholars have created and expanded taxonomies and coding systems to provide a base for storing, accessing and combining information from multiple domains. Second, the fields of mathematics, one of the foundational languages of all science, has evolved and expanded. Third, the educational infrastructure emerged over the past millennium to become a major agent in both controlling and promoting the codification and conservation of existing knowledge and the development and dissemination of new knowledge. Fourth, the academic and professional associations have had a major influence in defining the nature of disciplines and the conservation, accumulation and dissemination of epistemic specific knowledge. Finally, the institution of peer review and evaluation is embedded in most of the other institutions, as a rule of engagement and a unifying principle of knowledge development and preservation.

## Taxonomies

Taxonomy systems are at the foundation of our modern knowledge system. John McNeill argues that nomenclature, the principles and procedures governing the scientific naming and categorization of natural phenomena, is vital for all scientific communication.[18] In the area of biology, it has been said that an organism's name is the key to its literature.[19] Without internationally accepted and unambiguous naming, categorization and codification of knowledge, much of the information from past and current research could be irretrievable or at least difficult and costly to access. A range of systems have been developed over the millennia to replace the idiosyncratic, humanist structures of the pre-modernist period. These systems have ranged from the alphanumeric cataloguing systems, the Linnaean classification for living organisms, the periodic table in chemistry and an emerging international effort to standardize and define the array of more precise knowledge of our environment and society. Much of this effort can be traced to the eighteenth century, when more efficient storage devices such as textbooks and encyclopedias were developed and new and more powerful search engines such as indices and experts emerged.[20] The epistemic bases followed in tandem.

The starting point for the various codes of scientific knowledge varies from group to group. The earliest effort, in botany, began in 1753, and the most recent code, related to bacteriology, started anew in 1980. Although the earlier codes were initially the brainchild of individual scholars attempting to impose order in a field, increasingly after the 1890s these codes came under the aegis of international conventions.

The Linnaean System for biology has become the framework for classifying and naming living things based on a simple hierarchical structure, from most general to most specific.[21] The basic hierarchy as formulated by Linnaeus, moving from the general to specific, including separate classifications for empires, kingdoms (animal, vegetable, mineral), classes (mammals, birds, amphibians, fish, insects and worms), orders (each class may have many orders), genus, species and varieties. This hierarchical system still reflected the old medieval feudalistic worldview, where concepts like evolution were alien: the world and all its creatures were created once and simply awaited classification. Even though it is now generally accepted that evolution is proceeding within most kingdoms, this formula, as set out in the 10th edition of Linnaeus' *Systema Naturae*, published in 1758, is considered the foundation of all modern taxonomy. Over the intervening years, the structure of the taxonomic system has changed. Both empire and variety have ceased to be used and two new ranks were erected: *phylum* (or division in the case of *plants*) was added between kingdom and class, and family between order and genus, giving seven hierarchical ranks in all. This versatile arrangement is used to classify every living organism, living or extinct.

The degree of codification in the biological world has expanded to the point that there may now be too many competing taxonomies. Biology as a science is unusual in that organisms can be named according to five different codes of nomenclature: the International Code of Zoological Nomenclature which governs animals and the International Code of Botanical Nomenclature which governs plants, both of which began in the 1890s; the Bacteriological Code which began in the 1950s and was completely revised in 1980; the International Code of Nomenclature for Cultivated Plants which originated in 1953; and the Rules of Virus Classification and Nomenclature (which is working to convert the rules into codes).[22]

Each of the codes has a commission, a governing scientific body, a permanent secretariat and a process for handling both normal business and disputes. The International Code of Zoological Nomenclature, for example, was struck in 1895 and is currently administered by The International Commission on Zoological Nomenclature. The code's fundamental aim (similar to all the codes) is 'to provide the maximum universality and continuity in the scientific names of animals compatible with the freedom of scientists to classify all animals according to their taxonomic judgment'.[23] Each Code is administered by a Commission, supported by a full-time secretariat (the zoological group is located at the Natural History Museum in London), is mandated by the scientific membership of an international body (in the zoological case it is the International Union of Biological Sciences), and its members are elected by zoologists attending general assemblies of the international unions or other international congresses. In the case of the Zoological code, its current, fourth version has been produced in English, French, Catalonian, Chinese, Czech, German, Japanese, Russian, Spanish and Ukrainian. Over the years, several million species of animals have been recognized, and currently about 2000 new genus names and 15 000 new species names are being added to the literature every year. As one might expect, with so many names, problems do occur. The Commission makes nomenclature and taxonomic decisions in areas of conflict, after issuing a quarterly journal that solicits discussion by the zoological community.

The profusion of biological codes causes thorough nomenclatural investigations to be very time consuming and expensive.[24] A survey of United Kingdom taxonomists showed that some 20 per cent of their time in the early 1990s was devoted to nomenclatural work. This was the equivalent of 25–50 full-time taxonomic positions engaged solely on biological nomenclature, representing an annual expenditure of some US$1–2 million in the United Kingdom alone.[25] This led to efforts in the 1990s to develop a comprehensive

Biocode. So far the effort has not led to any resolution but it has helped to identify the problems and opportunities.

The periodic table for chemistry has had a somewhat different tale. Although several European scientists attempted to put the known chemical elements into some sort of order based on their various chemical and physical properties (such as melting point and atomic volume), the current version owes much to one created by Russian Dmitri Mendeleev, which arranged the elements by atomic mass. Henry Moseley subsequently improved the table by rearranging the elements according to atomic number. The Los Alamos scientists assert the periodic table is the most important chemistry reference.[26] The main benefit of the table is that is has allowed chemists to quantitatively explain the behavior of the elements and to predict the existence of yet undiscovered ones. There are 116 chemical elements whose discoveries have been confirmed; 95 can be found naturally on Earth and the rest have been produced in laboratories. The integrity of the periodic table and its related data is guarded by the International Union of Pure and Applied Chemistry, a worldwide organization composed of national chemical societies from 44 countries. IUPAC is recognized as the world authority on chemical nomenclature, terminology, standardized methods for measurement, atomic weights and other critically evaluated data.

A number of other taxonomic systems work to control the academic and research enterprise. In the first instance, granting councils, academic associations and journals all use taxonomic coding of research proposals and output. There does not appear to be any universal standard; rather, each organization, discipline or profession appears to develop their own coding for what constitutes a defined subject or methodology. Granting councils (such as the Social Science and Humanities Research Council of Canada) tend to organize their grant and research programs by a series of *sui generis* codes related to subdisciplines and research areas, undertaking budgeting, evaluation and management by discipline and area, which tends to concentrate researchers and institutional activities within a narrow range of the median. The academic associations and journals also codify people and research. The American Economics Association, for example, which publishes six journals and manages an electronic bibliography of global economics literature, categorizes its membership, its job search engine and its abstracting systems using its own subject descriptors. This de facto taxonomic system tends to focus attention and interest within narrow disciplinary boundaries, because it is used to manage both peer evaluation and classification of literature. Although there have been some efforts in many granting councils and academic associations to nurture more interdisciplinary work, the bulk of resources and energy is focused on conforming to narrowly defined disciplinary research embedded in the various taxonomies.

Meanwhile, the library sciences have gone a long way to making printed works more accessible. Perhaps the earliest and best known system is the Dewey Decimal Classification system, devised by Melvil Dewey in the 1870s. The strength of the Dewey decimal system is that it uses a coding of divisions and decimals to coordinate materials on the same subject and on related subjects to make items easier to find on the shelves by using a combination of letters and numbers. The Dewey system has ten main classes, which are each divided into ten divisions and then subdivided further. Each division becomes more specific; the more numbers, the more specific the subject. Now in its 22nd edition, the privately owned system is used mostly by public and school libraries.[27]

Taxonomies are like double-edged swords. They have great power to enhance and support knowledge management by enhancing searching and transmission of information and knowledge, but most systems have been subject to criticism. The more finely domains are defined, the less room there is for out-of-the-box thinking and investigation. In addition to the concern that the existing taxonomies narrow our range of research efforts, others have criticized them for unequally focusing on specific topics (such as the history of advanced industrial economies).

## The Language of Mathematics[28]

One of the most significant developments in the Enlightenment period was the rediscovery and development of mathematics as one of the foundational languages of discovery. Carl Gauss referred to mathematics as 'the Queen of the Sciences' because it is a body of knowledge justified exclusively by deductive reasoning, starting from axioms and definitions. The strength and adaptability of mathematics comes from the rigor imposed by the mathematical proof, which follows from axioms by means of systematic reasoning. This approach, formalized in Euclid's *Elements*, is completely analogous to Popper's falsifiability theory. The logical foundations of mathematical reasoning both ensures that mistaken 'theorems', based on fallible intuitions, are rejected and makes mathematics a powerful tool to investigate the unknown.

Every culture on earth has developed some mathematics. The Indian culture developed the decimal system and it migrated into Arabic and finally into its modern usage as early as the twelfth century. Mathematical theory, with roots in the classical civilizations of Greece, China, Egypt, Babylon, Japan and India, was internationalized when it was translated into Latin in the sixteenth century and became the mathematics of Western Europe and ultimately the world. One significant breakthrough was the replacement of the literary style of mathematics in the sixteenth century with the symbolic notation currently used; modern notation makes mathematics much easier, more accessible, more rigorous and more adaptable than the earlier, literary style.

Both applied and theoretical mathematics are important for understanding what we know. Although the earliest applied mathematics was used for very practical purposes, such as accounting or measuring, modern mathematics has been directed to a number of topics, including the study of quantity, structure, space and change.[29] Beginning in the 1870s, statistics began to develop as a related field of applied mathematics. Probability theory permits the description, analysis and prediction of phenomena where chance plays a role, such as when samples need to be taken or when there is a random element to a phenomenon. The combined stream of applied mathematical discovery has been instrumental in discovering and cataloging patterns, which makes it particularly relevant for the natural sciences, engineering, medicine and the social sciences (especially economics).

Pure mathematics, meanwhile, has been critical both in the advancement of applied mathematics and in the specific field of digital computation. Involving deductive reasoning, pure mathematics has led to the development of set theory and mathematical logic, which are both closely linked to the theories that underlie the development of the modern computer. When modern computers were conceived, several underlying theoretical concepts (such as computability, computational complexity and information theories) were shaped by mathematicians. More recently, numerical analysis has been heavily engaged in investigating computational methods for efficiently solving a range of problems involving large number sets, including the study of rounding errors or other sources of error in computation.

The evolution and content of the science of mathematics is governed by a number of structures. In the area of pure mathematics, peer evaluation by the various learned associations, academies, journals and departments of mathematics sustains the focus of the discipline on its epistemological roots. The use of applied mathematics is generally governed by a host discipline (either a natural, social or applied science community, involving various academic trappings, or a professional association, such as the professional engineering societies). When mathematics crosses into physical products, other organizations become involved. The various standards setting bodies define and prescribe the appropriate standards for the use of statistics and other measurement techniques in the safety and quality systems. For example, the International Electrotechnical Commission (IEC) is the international standards organization dealing with electrical, electronic and related technologies which has responsibility for dealing with the rounding problem (decimal floating-point computer algebra systems, and certain big number systems, might give either the answer to $3 \times 1/3$ as either 1 or 0.999 . . .). The IEC, with 65 member countries and 69 affiliate countries, sets standards for all electrotechnologies including energy production and distribution, electronics, magnetics and electromagnetics, electroacoustics, multimedia and telecommunication, as well as associated general disciplines such as terminology and symbols, electromagnetic compatibility, measurement and performance, dependability, design and development, safety and the environment.

As with all governing systems, mathematics creates winners and losers. Kenneth Boulding asserts that as 'mathematics is a tool or a language' it inevitably 'involves abstraction (orderly loss of information)'.[30] This process of abstracting from the complexity of natural phenomena is a critical part of the scientific method, but also involves a loss of 'richness' of information. Although the losses may be justified in terms of the results that can be gained from the deductive process, one cannot ignore that some perhaps important variables may be lost in the process. Values, which tend not to be quantifiable in any ordinal or cardinal sense, are particularly at risk in a mathematically defined world.

## Universities and Public Research Institutions

The university, as an institution of higher education and research, has become a cornerstone of the knowledge economy. Virtually all scholars of the modern, industrial economy accept that universities play a critical role as members of the triple helix or the research community, acting as repositories for existing knowledge, as purveyors of knowledge through education and as creators of new knowledge through research.

Universities emerged in a wide range of structures in many parts of the world over the past three millennia. Although institutions of higher learning were found scattered through many of the ancient civilizations, including China, Egypt and India, the basis for the modern university can be traced back to the Academy, founded by Plato near Athens in 387 BC. The modern university can be traced to the medieval institutions that began to emerge from cathedral schools or monasteries beginning in the ninth century in Constantinople and Salerno and extending over the succeeding years to Bologna (1088), Paris (*c.* 1100), Oxford (*c.* 1100) and Cambridge (1209).

Over the intervening period, the nature, structure and scope of these institutions has changed widely. In the early, classical period, universities were about philosophy, mathematics and gymnastics. During the medieval period grammar, rhetoric, logic, arithmetic, geometry, music, and astronomy were added to the modern languages. The Reformation worked to sever the connection between education and moral pursuits, and increasingly during the Enlightenment period differentiated disciplines emerged. By the turn of the nineteenth century much of the academy had become compartmentalized, with increasingly narrowly defined disciplines organized into departments and colleges – the social sciences, for instance, divided into economics, political science, sociology, anthropology and psychology. Meanwhile, many universities became home to a wide array of professional colleges, such as agriculture, commerce, dentistry, engineering, law, medicine and pharmacy.

The wide range of activities now encompassed within universities makes it hard to determine how many universities exist in the world today. The International Association of Universities uses UNESCO data to maintain the World Higher Education Database, which listed almost 9760 university level institutions in 184 countries and territories in 2005.[31] It also identified some 8000 non-university level institutions of higher education.

These institutions are governed both externally and internally. In the first instance, most universities are generally established by statute or charter, granted by the responsible government in the region. Most jurisdictions also grant universities the monopoly on the granting of degrees. Once an institution is set, the chartering agency often loses some control and much of the governing is internal to the university. Lawrence DeNardis, president of the University of New Haven, suggests that the responsibilities divide between the faculty and the administration: curriculum, faculty status and standards of student competence are responsibilities of the faculty; institutional budgeting, staff salaries and physical and fiscal resources are the responsibility of the administration; creation or elimination of academic programs is a joint responsibility; and on some other matters, responsibility is not clear or is disputed.[32] In addition to the internal checks and balances of the academy, external accreditation agencies undertake extensive ongoing evaluation and assessment. Graduate programs and professional schools such as law, nursing, medicine and engineering almost always have rigorous discipline-specific accreditation procedures. If a deficiency is found, a university, department, college or program will have a finite period of time to remedy the problem before losing accreditation. In most cases, the loss of accreditation would jeopardize the future of the program, as both students and money would dry up. Each country operates their system somewhat differently. In the United States, accreditation is carried out by private, non-profit, specific-purpose organizations. The US Council for Higher Education Accreditation, for example, works with 60 accreditation bodies that certify over 7000 degree-granting and non-degree-granting institutions and over 17 000 programs. Accreditation in countries other than the US is normally handled by ministries of education or other government entities. In the UK, the Quality Assurance Agency for Higher Education works with devolved education authorities in England, Wales, Scotland and Northern Ireland, and the Japan University Accreditation Association, Inc. is an independent organization of national, public and private universities. Canada, in contrast, has no formal system of institutional accreditation; institutions tend to be vetted by their provincial governments and adhere to codes related to the membership in the Association of Universities and Community Colleges.

Private postsecondary schools (not including universities) tend to be regulated by provincial or state governing bodies, approving schools to operate and ensuring that training is relevant and practical. In those cases, the regulators may have an ongoing role. California currently has more than 2000 private postsecondary schools under its jurisdiction. A bureau of the Department of Consumer Affairs, California's consumer advocate in state government, establishes standards of competency for more than 200 professions and occupations and administers the regulatory functions of licensing, examination and enforcement for those professions.

## Journals

The advent of the print culture in the 1400s was a critical event in the governing of new knowledge.[33] The manuscript culture, where texts were hand-copied, significantly impeded the dissemination of knowledge and at times led to its degradation, as errors crept into successive generations of hand-transcribed texts. The print culture allowed wider dissemination of a larger variety of traditional and new products (such as catechisms, calendars, guide-books and manuals). Over time, a wide variety of new products emerged that worked to standardize knowledge, including textbooks, journals, technical encyclopedias and various engineering and applied science manuals.

The new technology also increased pressure for standardization of the style and usage of the printed work, including spelling, grammar, printing styles (such as gothic and roman), layout, binding and structure of printed works (such as title pages, tables of contents, indexes and bibliographies). Indexing, cataloguing and cross-referencing emerged. Printed works could increasingly be disseminated more easily, allowing more critical comment and debate about the nature of ideas. This, combined with the innovation of the scientific journal in 1665 (credited to Jean Baptiste Colbert, contrôleur général under King Louis XIV of France), created the foundation for the academic discourse that has emerged in the past 350 years. Overall, there was a culture of improvement rather than degradation of the texts, as printers fixed errors that had crept into hand-transcribed texts. The printing revolution also led to the emergence of the new property and power structure embodied in copyright. Initially designed to control the spread of works that could challenge the state (copyright was initially granted to publishers, not authors, and they were held liable for any seditious or blasphemous works and for controlling the dissemination of texts), increasingly it became a tool in support of diversity and individual enterprise (copyright was granted to authors in 1710 in the UK and in 1794 in France).

There are currently about 50 000 peer-reviewed journals being published by approximately 17 500 scientific scholarly societies. These societies, largely through their management of their scholarly journals, do the bulk of the work in governing their disciplinary epistemology. Although the journals are extensions of the academic enterprise, a large portion of the academic press now operates as for-profit enterprises that seek to balance academic integrity with profit. Many of the biggest and best known journals are now the sole property of commercial ventures. Nature Publishing Group, the scientific publishing arm of Macmillan Publishers Ltd, produces 20 related life-science journals, including the highly cited *Nature*; Emerald, established in 1967, produces more than 150 journals and library information services for engineering, applied science and technology; Kluwer Law, a division of the Wolters Kluwer publishing group, developed more than 50 legal and social science journals and recently purchased the Thomson/Chapman & Hall science and technology program; and Elsevier operates more than 2000 journals across a wide range of subjects and areas, including the *Lancet Journal*. Many journals remain the purview of their epistemic communities, including the largest paid circulation journal *Science*, published by the American Association for the Advancement of Science, an international non-profit organization with 262 affiliated societies and academies of science, and the *New England Journal of Medicine* and the *Journal of the American Medical Association*, which are governed by Massachusetts and US medical societies. Regardless of who owns and operates the journals, there is increasing pressure for them to become more commercial and popular. *Nature*, for example, will issue a press release related to 'exceptionally interesting papers' and is proud to point out that it is subscribed to by more than 2000 journalists and media organizations worldwide. It markets this to authors, reminding them that those papers that are featured will receive 'maximum exposure for their work in the world's most important newspapers, magazines, radio and television channels'. Even those journals that are owned and operated by the epistemic communities are entering the business of promoting their work.

There has been significant debate in recent years about the role of journals in normalizing and preserving knowledge. They certainly occupy a dominant position in the publishing world. A recent report showed that the more than 2000 scientific, technological and medical publishers around the world offer about 16 000 periodic journals which publish 1.2 million articles annually, and the number of articles rose by about 20 per cent over the 1998–2003 period.[34] Three contentious issues have dominated.

First, even though supply is rising, there is increasing concern that the rising cost of journals is leading many universities and research institutions to reduce their subscriptions to journals. The UK House of Commons Science and Technology committee found recently that the average price of an academic journal in Britain rose 58 per cent between 1998 and 2003, compared to an 11 per cent rise in the overall consumer price index. A particular concern is that the rising prices may not reflect the costs of publication but rather the profits. *The Economist* reported that Reed Elsevier, a UK publisher owning the Dutch subsidiary Elsevier, had operating profits of about 34 per cent in recent years.

Contrasted against these concerns about cost and profit is the question of access. The granting agencies are expressing increasing concern about the pricing practices of the journals. Given that much of the research published in journals is funded by public or philanthropic agencies, they are concerned about the ultimate use of the knowledge generated. Many journals are reaping much of their profits from exploiting their copyright on articles for extended periods of time after publication.

A second concern relates to management and use of the vast databanks of journals. One contentious issue is how various weighting schemes can and might influence research. A number of organizations undertake to rank individual journals for their 'impact', usually defined as the rate at which articles published in a journal are cited by others journal articles. The higher the citation rate, the more impact a journal is supposed to have. The Institute for Scientific Information (ISI) has for more than 50 years been developing a bibliometric database of research. Its three subject indexes – the Science Citation Index, the Social Science Citation Index and the Arts and Humanities Citation Index – combined have 35 million entries from 14 800 journals in 230 disciplines, reaching as far back as 1900 for the sciences (to 1956 for the social sciences and 1975 for the arts and humanities). This system has the ability to develop bibliometric pathways, impact factors and a variety of other measures of performance. Although it provides a valuable tool for researchers to find other related materials (and to see how their work is being used), it also is being used to evaluate journals and individual researchers. The Arts and Humanities Research Board in the UK, for instance, has been consulting with academia to compile lists of the top ten 'most important and significant' journals in each of ten subject areas, in response to a UK government request for 'quantitative indicators' of research quality to justify its £70 million research budget.[35] The board is seeking to measure the impact of the work it funds by analyzing the numbers of articles by UK authors in listed journals. Some have expressed concern that an official hierarchy of journals could kill emerging disciplines and publications and could either distort or harm the careers of academics. This is particularly a concern for those scholars who operate in interdisciplinary areas, because their papers, though frequently influential and at times transformative, are seldom in high-impact publications.

The third main change is the revolution taking place in scientific publishing. The advent of electronic publishing is changing the economics and has the potential to change the management of knowledge. For traditional, paper-based journals, the main cost of operating is the printing, publishing and mailing of the journals. With the emergence of the Internet and various new protocols and programs for transferring and storing information, there is now the real possibility of making articles more accessible to more people at a lower cost than ever before. The so-called 'open access' initiative is designed to provide free access to published peer-reviewed research articles, either immediately or after a limited embargo period (set to six months to a year to enable the publisher to cover the costs of peer-review coordination, editing and archiving). There is increasing pressure from funding sources for more open access. In October 2003, the leading research associations of Germany, France and Switzerland concluded the 'Berlin Declaration' which called for free access to research findings and the Max Planck Society has changed its employment contracts to require staff to return copyright of their work to the society. In the US, the House of Representatives passed in August 2004 a provision that would require results of studies funded by the NIH (estimated to be up to 50 000 papers per year) to be deposited into the online National Library of Medicine's archive called PubMed Central within six months of publication in any journal.

A recent scan suggests as many as 12 000 journals may have some measure of open access. Some are purpose-built open access, e-journals. BioMed Central, for instance, is a relatively new, independent publishing house that operates more than 130 disciplinary, peer-reviewed electronic journals for medical scholars and researchers to publish with full and immediate open access.[36] BioMed charges between US$585 to US$1680 for each article to defray the costs of operating the system (this may seem high but some traditional paper journals are in the practice of levying page charges of as much as US$1000). So far, more than 23 national granting councils and charitable research trusts in 19 countries have agreed to allow their grant recipients to use some of their funds to pay the charges.

Many traditional journals are similarly entering this field. The *Proceedings of the National Academy of Sciences* (*PNAS*), for example, recently has begun an open access program, where it made all content freely accessible without charge from the first volume in 1915 to articles published just six months previously. The articles were put on their website[37] and on PubMed Central.[38] The journal also made all of its content free online immediately upon publication to 145 developing countries. Furthermore, the journal changed its copyright and permissions policy to allow authors to post the PDF of their article on their web site, to post and update preprints and to post webcasts for education purposes, and anyone is now allowed to reuse original figures and tables published in PNAS for educational purposes without having to request permission. Finally, the journal has undertaken an experimental open access option for PNAS authors, where authors may pay a surcharge of $1000 to make their article freely available online immediately upon publication.

This democratization and opening up of the academic press is far from over. The impact on the management of knowledge is unclear. Many scholars report that the easier and cheaper access to journals – which can be searched from a researcher's desk rather than by going to a library to use periodical indices to find and retrieve paper copies – is changing the way they do research. Many now report using a larger number, wider array and greater diversity of journals in their research, than in earlier times. Although this is undoubtedly valuable, there is still some uncertainty about the quality of much of the on-line material. Many of the open access journals have highly qualified and well respected editors and editorial boards; others may not. So far there is limited quality assurance undertaken on e-journals. This may be in the process of changing. Some of these new open access journals have been assessed by ISI for listing in their various citation indices and early indications are that because of their ease of access and the ability to quickly search their content electronically they are receiving equal or higher citation rates than some of the top ranked journals in their fields.[39]

## Peer Review

Peer review is inextricably linked to the emergence and structure of epistemic communities, learned journals and universities. The concept of peer review, implicitly embedded in Boyle's conception of science, only really took hold as a governing rule in the academic world in the nineteenth century. Before then, scientists, scholars and inventors would share their work for review and comment with their peers, but it was more of a principle than a rule. There was no formal peer evaluation of the conception of the projects, nor was there a formal review of the final results. Rather, open disclosure and repeated experimentation were the means by which verifiable research was normalized and inaccurate or flawed knowledge was removed from the canon. Beginning in the nineteenth century and especially after the Second World War, peer evaluation of both research proposals and research findings increasingly became the rule. This was undoubtedly due, in part, to the increased role for public funding of research (and the imperative to show value for money spent by the public sector) and the greater volume of research seeking to be published (and a need to develop some system for choosing what to publish). Even though the number of journals has ballooned in the past generation, the number of top ranked and highly cited journals remains small in any discipline. Hence, as the volume of potential articles went up, the disciplines needed a means to select the best of the articles for publication in the best journals.

Peer review is generally intended to delimit the 'scope of knowledge claims that will ultimately appear in print'.[40] It generally involves some process where editors of journals (or administrators of granting programs) solicit opinions of knowledgeable peers about the work others have done (or propose to do). Most reviews are structured in a way to answer a few basic questions. Generally, reviewers of candidate journal articles are asked to comment on the quality and novelty of the problem under investigation, the theoretical underpinnings for the work, the methodology developed, adapted or adopted, the data used and the conclusions reached, in addition to their opinion on the readability of the work. In this sense, the review is intended to address Polanyi's three merits of plausibility, value and originality. Peer review of grant proposals is analogous, except that most granting agencies also want the reviewer to comment on the merits of the researcher and not simply on the research. Reviewers are asked to assess whether the project proponent has the background for the work and has demonstrated capacity to complete work of the nature proposed. In that way, review of grants fails the test of anonymity proposed by Boyle and may, at times, lead to conflicts of interest or personalities.

Peer evaluation is both a powerful and disputatious tool. Most peer review is intended to operationalize Boyle's dictum that we should assess the merit of the findings rather than the merit of the scholar. Hence, most peer structures attempt to create some independence between the reviewer and the reviewed. Many journals operate double-blind reviews, where both the author and the reviewer are anonymous. Although this offers the greatest hope for an impartial review, in many disciplines there are so few potential authors and reviewers on specific topics that it is relatively easy to figure out who is writing or reviewing a piece. In other instances, only the reviewer may be undisclosed, or neither may be kept anonymous. Regardless of the degree of transparency, it is important to remember that the reviewers do not decide on articles, they only advise the editor, or editorial board, of the merit of an article. The final choice is left up the editor. In some cases the journals will simply use the editor, or the editorial board, as the reviewers.

Although peer review is the currency of academic and scholarly activity, it is neither the only way to identify meritorious work nor is it fail proof. Critics of peer review gleefully note that some of the twentieth century's greatest research findings were not reported in peer-reviewed journal articles. Albert Einstein's papers on the theory of relativity in the 1905 issue of *Annalen der Physik* were not peer reviewed – the editor, Max Planck, simply published them on his say-so. Similarly, Watson and Crick's 1951 paper on the structure on DNA in *Nature* was not sent for review – the editor asserted the correctness of the work was 'self evident' and there was no referee working in the field who could have kept it secret prior to publishing. There are other examples of papers produced by future Nobel Prize winners that were not reviewed. Furthermore, peer reviewers cannot be expected to identify all cases of academic fraud, dishonesty or even simple error. There have been a range of well publicized examples of papers accepted for publication that contained results that either were falsified or tampered with. A recent example is the case of Hwang Woo Suk, from the Seoul National University, who published high-profile articles in *Science* and *Nature* about creating 11 human embryonic stem cell colonies. A Seoul National University investigation found that at least nine of 11 colonies of stem cells either never existed or were never proved to be stem cells. Similarly, a recent *Nature* review article noted that the on-line, open-access, non-reviewed Wikipedia encyclopedia made about the same number and scale of errors about scientific topics as the prestigious, formally edited and reviewed Encyclopedia Britannica.[41]

Thus, peer review is an important process delimiting our academic and scientific knowledge, but it is not without flaws. It is particularly stressed when dealing with contested science, especially when it relates to transformative innovations. In those cases it is often hard to find reviewers, editors at times may overrule their reviewers and industry, the state and various publics are engaged in examining, assessing and valuing the knowledge claims.

## GOVERNING CONTESTED KNOWLEDGE

Transformative innovation is fundamentally about the introduction of new, revolutionary science that changes the way we see our circumstances and our opportunities. As with other changes, there are always vested interests, values and circumstances that will lead to reactions against new ideas. In some cases the forces of conservatism are justified, in that the new candidate knowledge either is incorrectly formulated or simply does not generate any new value to the economy or society. But, as discussed in Chapters 2 and 3, the value and impact of any new transformative knowledge is delineated only through extensive trial and error. Thus, there is a significant problem in normalizing contested, revolutionary knowledge. To generate optimal value, it needs to be normalized into a paradigm but, unlike for normal knowledge where there are longstanding and well developed institutions that either individually or collectively delimit the nature of our knowledge, contested knowledge does not have any well established set of institutions. Rather, an array of ad hoc, often purpose-built, approaches are used.

Every modern transformative technological revolution has led to a period of contested science, where some knowledge claims are readily accepted, some are rejected out of hand and some live on in the hearts and minds of people, generating heated debate and confusion about what truth to believe. The area of agricultural biotechnology offers three global case studies of how new, dramatic (and some would say antithetical) knowledge claims were introduced, adjudicated and at least partially resolved. When biotechnology entered the food market in 1994, there may have been a few people and groups with latent concerns, but expert and public opinion ranged from enthusiastic to indifferent. That situation continued until 1998, when two studies shook the foundations of the rapidly expanding global agricultural biotechnology industry. In 1998 a study in the UK raised concerns about the safety of genetically modified insecticidal potatoes and in 1999 a study in the US suggested *Bt* corn could have unintended adverse effects in the environment. The one-two punch of the two studies unleashed a firestorm of claims, counterclaims and processes to mediate the contested claims. Then in 2001, a study was released claiming that GM traits had introgressed into landrace corn and teosinte in Mexico, which triggered an international challenge to the technology in the context of biosafety and traditional knowledge. Each of the three triggering studies was done by credible scientists working in what they thought was 'normal' science. Although there has been extensive debate about the merit of the science underlying the claims, the most important consideration for this study is how the contest over the knowledge claims was governed.

## The Pusztai Controversy About the Safety of GM Potatoes and Other Foods

The first case represents the challenges of dealing with what Gibbons calls 'post-normal' or mode-2 science.[42] In August 1998, Dr Arpad Pusztai, a senior research scientist at the Rowett Research Institute in Aberdeen, Scotland, went on the BBC *Panorama* television program in England and announced he would not choose to eat GM foods, noting that his research (then unpublished) showed that rats fed on potatoes genetically modified to contain lectins from the snowdrop plant – a naturally occurring insecticide – suffered damaged immune systems and stunted growth of vital organs after a period corresponding to ten years of human life.[43] This was the first scientific evidence that suggested GM crops could be hazardous and contrasted sharply to the science of the day. Within four days of his television appearance, the Rowett Institute has suspended him and directed him not to talk about his findings. Although there are many comments attributed to the Rowett Institute and to the supporters of Dr Pusztai about the cause for the response, the formal reason for his suspension was that he had presented the results publicly before they had been peer reviewed by other scientists, as normally required by the Rowett Institute. Over the ensuing months, Rowett assembled a scientific committee to review Pusztai's study; it concluded there were important deficiencies in the work but that any errors did not represent scientific fraud. In response, Dr Pusztai sent his research to 24 scientists in different countries; this group concluded that Rowett's scientific committee erred, finding instead that the research was of good quality and the conclusions were justified.

The leading scientific journal *The Lancet* then entered the fray. The editor of the journal invited Dr. Pusztai to submit the paper,[44] sent it for peer review (in this case to six reviewers rather than the normal two) and then, even though the majority of the reviewers recommended the paper be rejected, published it in the 'Research Letters' section of the journal (about eight months after the August television broadcast). The editor of *The Lancet* also unusually offered a lengthy explanation of the reasons for the decision to publish, arguing that publication of the paper did not certify the scientific merit of the work but rather was a response to the level of public anxiety about the issue. He quoted one of the reviewers as saying that although the research was flawed, he would like to see it in the public domain so that fellow scientists could judge for themselves; failure to do so might hint at a conspiracy to suppress information. Although well intentioned, *The Lancet* publication even further exacerbated the issue. The authors and their supporters took publication as vindication while the opponents asserted publication raised the potential that research in journals could be driven by public debate rather than 'good' science.

Following this, a wide range of collegial processes was engaged in many countries around the world to review either the Pusztai results or the more general question of the safety of genetically modified foods. The Royal Society in the UK, for example, undertook a specific review of Pusztai's results, resolving that they were inconclusive and, in some ways, flawed. One implication drawn out by Dr Pusztai was that substances in GM foods that have a slow-acting effect would not likely be detected because the prevailing regulations did not require long-term safety testing. If correct, this would challenge the operation of the regulatory regimes in most countries around the world. This concern triggered a succession of national and international reviews of the safety and regulation of GM foods.

In the US, the National Academy of Science and the US President's Council of Biotechnology Ethics undertook separate reviews of the technology while the regulators (FDA, EPA and APHIS) all reviewed their procedures and the Pew Institute created an Initiative on Food and Biotechnology to undertake ongoing review of emerging issues. In Canada, the federal government both invited the Royal Society of Canada to empanel a group to review the science underlying the regulation of GM foods and established the Canadian Biotechnology Advisory Committee, an expert committee initially of 21 scientists, social scientists and ethicists, to review the operation of the regulatory regime and the communication of risks with citizens. In the UK, the Royal Society empanelled an independent working group of ten scholars to review the issue of GM plants for food use and develop a statement that was reviewed and endorsed by the council of The Royal Society.[45] The working group issued a call for evidence: 13 organizations and eight individuals responded with written briefs and ten additional individuals (from industry, government, academia, and NGOs) met with the committee for oral evidence sessions. Their final report reviewed the current status of GM foods, reviewed the scientific evidence of hazards and recommended a range of changes to the regulatory system in the UK. At about the same time, the UK government established the Advisory Committee on Releases into the Environment (ACRE) and the Nuffield Council on Bioethics undertook a comprehensive ethical review of the technology and its impacts. The EU, in contrast, enacted a de facto moratorium on further crop approvals (having approved 18 GM crops up to that point), undertook extensive consultations about structuring a new set of European regulations for new crops and commissioned more than 60 separate research projects to review the human health, biosafety, economic and social impacts of the technology. The New Zealand Government established a full-scale, multi-year Royal Commission On Genetic Modification, Australia engaged in extensive research and national and state consultations in the lead-up to developing an Australian Gene Technology Advisor. Meanwhile, new models of citizen engagement were tested. Citizens' juries were developed in Denmark and spread widely through the EU and into Canada, with more than 25 juries empanelled by 2005.

Because the science is international, the effort to debate the science went global as well. The OECD was invited by the Group of Eight (G8) leaders at their summit in Cologne in June 1999 to 'undertake a study of the implications of biotechnology and other aspects of food safety'. In addition to empanelling groups of scientists to review specific aspects of the science and regulatory system, they also convened an international conference on food safety, hosted by the UK government in Edinburgh on 28 February to 1 March 2000 and chaired by Sir John Krebs, Professor of Zoology at Oxford University and the chairman designate of the future UK Food Standards Agency. The conference brought together 400 participants from more than 40 countries, including representatives of governments, industry and civil society organizations, such as Greenpeace International, Friends of the Earth and GeneWatch. The OECD submitted its final report to the G8 leaders' summit at Okinawa, Japan, in July 2000, but it didn't stop the debate. There has been a succession of international reviews under the aegis of the FAO, WHO, the World Bank and many of the international learned societies (culminating in a statement by the United Nations Development Programme in Agenda 21 in 2004, asserting that cautious, selective use of the technology could benefit humankind), all attempting to provide greater clarity to the science and its impacts.[46]

In retrospect, the Pusztai controversy offers a number of lessons about governing contested knowledge. Right from the beginning, the debate exhibited post-modern attributes. Dr Pusztai released his results to the public before peer review, he made claims (mainly that GM food is unsafe) that exceeded the norms of the scientific method and he engaged in a personal campaign of validating his views by reference to authority rather than by repeated experimentation. Second, although many of the subsequent review processes generated lively debate about the nature of science and its use, they failed to come up with a clear process for judging knowledge claims in the area of contested science. A particularly frustrating aspect of the discussions is that when non-scientists were invited into the processes, scientists tended to drop their normal scientific decorum and present a mélange of both normal science and their musings about potential concerns (such as their next research issue), without distinguishing between them. Hence, it is seldom clear what is normal science and what is emerging (hypothetical, experiential or speculative) science. As a result, those citizens and consumers who engage in trying to sort out the state of the science, its risks, its benefits and its uncertainties are often thwarted. Third, the tenor of the 'post-normal' debate was disturbing. In a direct throw-back to earlier times, the debate became personalized; instead of playing the ball (by criticizing the science), many participants played the man (criticizing Pusztai, the Editor of *The Lancet* or one of those who took sides in the controversy). This personalization of scientific debate, though perhaps understandable given the perceived interests at stake, does little to clarify the claims of contested knowledge.

## The Losey Controversy About the Environmental Impacts of GM Corn

The second controversy represents an example of how normal science deals with controversy. In contrast to the Pusztai controversy, this case was triggered more conventionally. Dr John Losey of Cornell University, collaborating with two colleagues, produced an article for *Nature* in 1999[47] that claimed that pollen from transgenetically modified corn (containing a gene from a soil bacterium that expresses for *Bacillus thuringiensis*, a naturally occurring insecticide) could harm Monarch larvae. Losey and his team demonstrated in laboratory experiments that Monarch butterflies at the larval stage that ingest pollen from transgenic corn expressing a *Bt* toxin suffered increased mortality or delayed development compared with exposure to non-*Bt* corn pollen.

This research, similar to that done by Pusztai, generated significant scientific and public concern and debate. Unlike the Pusztai case, however, one might conclude that the scientific considerations were handled expeditiously and cleanly within the realm of normal science. The original article, for instance, in line with the scientific method, made narrow, delimited claims, noting the results, based on experiments in the laboratory, were preliminary and would need further investigation. The peer review of the article by *Nature* went according to normal procedures, and although the article triggered significant debate and criticism in the academic community, there was no claim that the paper should not have been published. Rather, the focus has been on examining and validating or refuting the science. Although many disputed Losey's results, that early research did much to stimulate needed research in this field by highlighting the potential problem.

None of this implies that the debate about the science was mild and commonplace. Rather, there were heated criticisms of the Losey results coming from many quarters. Critics pointed to various methodological concerns – the lack of controls in the tests, the extrapolation from laboratory-based studies to the field, and the question of density-dependent factors that would affect exposure of Monarch butterfly populations in the wild to the *Bt*-corn pollen. Although experimental trials submitted to the regulator showed that *Bt*-corn was not toxic to lady beetles, green lacewing larvae, spiders, minute pirate bugs or parasitic wasps, *Bt*-corn was thought by scientists and regulators not to be a problem for Monarch butterflies because the preferred food plant of larvae, milkweed, is infrequent in corn fields and exposure was expected to be minimal because *Bt* is inactivated by ultraviolet light and drought conditions. Furthermore, the current practice of spraying broad-spectrum insecticides for corn pests was judged to a significantly higher risk to Monarch populations than that posed by *Bt*.[48] Nevertheless, scientists readily admitted that the full potential for exposure of non-target Lepidoptera to toxic pollen was not critically examined in field trials. Rather than defend their work, Losey and his co-authors readily acknowledged many of the criticisms of their study and actually engaged in some of the follow-on research to clarify the knowledge.

Shortly after the Losey study was published, the Agricultural Biotechnology Stewardship Working Group in the US funded 17 studies to examine the risk of *Bt*-corn to Monarch butterflies[49] while in Canada the CFIA, Environment Canada and the University of Guelph set up a study to examine the potential impacts in Ontario. Those projects and other academic and contracted research looked at a wide range of potential adverse effects on non-target insects. The results of the studies are less important for our purposes than the fact that the process was largely collegial and worked according to the strictures of normal science. Research was either commissioned or initiated by independent scholars, studies were completed and results were peer reviewed and published in a variety of journals or reports. Although there was significant variance in some of the results, the work proceeded along the lines of normal problem solving. In addition to delimiting the knowledge about impacts on Monarch larvae, the research expanded to address a variety of issues related to insecticidal strategies, including the role of honey bees in transmitting pollen, the relationship of *Bt* to beneficial aphids and the potential for roots or post-harvest plant residue to exude toxins that harm non-target organisms in the soil (such as earthworms, various microbes, protozoa and nematodes). As the research accumulated, scientists, regulators, agronomists and farmers gained a greater understanding of the nature of the interactions of intensive agriculture and the ambient environment.

Even though the scientific process proceeded along normal lines, the nature of Losey's knowledge claims and the actions of the journal *Nature* in promoting the original journal article triggered an extensive public debate. Many of the environmental NGOs and engaged scientists jumped to speculative and premature conclusions about the effects of *Bt* pollen on non-target organisms and took as given the original findings. The subsequent public debate about the efficacy and environmental safety of the technology exemplifies a general public misunderstanding of how science is done. Most critics and opponents of biotechnology took on board the results of Losey's original study but ignored all of the follow-on research that modified the conclusions. To this day one can read definitive statements on the web, based on Losey's original article, asserting that *Bt* corn is a major threat to Monarchs. If anything, the conclusion of the package of studies is that although the risk remains, the actual threat is small and, in some cases, may be lower than the conventional agronomic practices that it replaces (specifically intensive aerial or ground spraying of infested crops).

The Losey controversy offers a few lessons about governing normal knowledge in a post-modern world. Although the science has been governed by the institutions of normal knowledge (especially universities, journals and peer review), the knock-on debate in the public realm has been more difficult. In a way similar to the Pusztai case, the public debate was messy. Those few scientists who engaged in the public discussion faced a barrage of criticism from their peers. Few scientists could claim comprehensive capacity in the subject area, so many were open to challenge on the basis that they were engaged beyond their epistemic competence. Some of the processes initiated in response to the Pusztai controversy ended up being drawn into the parallel discussion on the environmental impact of GM crops.

## The Quist and Chapela Controversy About Biosafety Hazards of GM Corn

A third controversy emerged in 2001, when plant biologists David Quist and Ignacio Chapela reported in *Nature* that in 2000 they had detected transgenes in local maize landrace varieties in the mountains of Oaxaca, Mexico.[50] The two UC Berkeley scientists reported that they had used PCR-based methods and a dot-blot DNA hybridization technique on a pooled kernel sample of Mexican corn from the region and that their tests detected the presence of the 35S cauliflower mosaic virus promoter sequence, the promoter used in all GM corn varieties commercially available in the US. They also used PCR-based methods to test for the nopaline synthase (NOS) terminator sequence, also used in the GM maize varieties in the commercial trade.

Once the results were released in *Nature*, critics challenged the findings on the basis that inverse PCR analysis, as undertaken by Quist and Chapela, was especially prone to artifacts and false positives. *Nature* reviewed its publication decision and took the unusual step of withdrawing the paper on the grounds that it did not meet the peer-review standards of the journal.

In absence of definitive scientific evidence, various processes were engaged. First, a number of scientific teams attempted to validate the claim that GM traits had introgressed into landraces. A number of other field studies were undertaken in an effort to confirm or refute the results, but none was successfully peer reviewed until 2005. The first peer-reviewed study found no evidence of any transgenes in the crop or in government storage facilities in either the 2003 or 2004 crop years.[51]

Meanwhile, the public controversy continued to build both in Mexico and throughout the NGO community. In April 2002, the Commission for Environmental Cooperation, a creation of the NAFTA side deals in 1994, was petitioned by 21 indigenous communities of Oaxaca and three Mexican environmental groups, Greenpeace México, the Mexican Center for Environmental Law (Centro Mexicano de Derecho Ambiental) and the Union of Mexican Environmental Groups (and eventually supported by more than 90 letters from organizations and institutions throughout the three NAFTA countries) to undertake an investigation under Chapter 13 of the agreement. This issue was considered of great potential environmental importance, given that Mexico is a center of origin and diversity for maize and that maize is so intrinsically linked to Mexican culture, especially that of Mexican indigenous groups. The Advisory Group on Maize and Biodiversity, involving 16 participants from industry, academia and NGOs in Canada, the US, Mexico and the UK, committed to using objective science and the participation of stakeholders to transparently develop a report. The group, in collaboration with the CEC Secretariat, commissioned a set of background chapters from a range of academics and practitioners about specific topics. The papers were then peer reviewed and published in a compendium to the final report. The Secretariat and the Group also convened an open, public symposium in Oaxaca where summaries of the chapters were presented and discussed. The Group then met and formulated a set of key findings and recommendations based on the background chapters and their peer reviews, the professional expertise of the advisory group members themselves, comments received at the public symposium and subsequently, and the comments of the parties to the North American Agreement on Environmental Cooperation (the governments of Canada, US and Mexico).

At one level, the CEC process worked to delineate the contested knowledge claims. The advisory group concluded that gene flow between landraces of maize – as well as between landraces and modern varieties – has been demonstrated to occur experimentally and descriptively and that, though the evidence of transgenes entering landraces of maize in Mexico has not been peer reviewed, there is no doubt that transgenes will spread in Mexican maize and that they were likely present at the time. The advisory group did not offer any insights into how the genes entered the country (as at that time GM corn was not to be imported for cultivation) and expressed no decisive opinion on whether novel alleles (transgenic or not) would eventually increase or decrease in frequency or whether any novel alleles would appreciably affect the biodiversity of the landraces.[52] Beyond that, the advisory group noted that 'high levels of poverty, dependence upon agriculture by large populations for income and food security, and a significant indigenous population distinguishes rural Mexico from that of Canada and the United States'. The endemic 'rural crisis' in Mexico of poverty, migration, and dislocation has:

> . . . become intertwined in the debate over the impacts of the presence of transgenes in Mexican landraces. Care needs to be taken by decision makers to recognize the impact of broader issues upon the views and interests of proponents and opponents of transgenic maize in Mexico. Thus, the controversy over GM maize may either reflect important underlying positions or be exploited for political purposes by opponents and proponents of GM crops.[53]

The advisory group went on to recommend a range of scientific, regulatory, industrial and social responses to the presence, or impending presence, of transgenes in landraces in Mexico, but few of the recommendations satisfied those concerned about the issue. The process withered away, as the final report was kept under wraps until the Friday after the US Presidential Election in 2004, when it was unceremoniously released without comment by the member countries.

This case offers a few additional lessons about governing contested knowledge in a post-modern world. First, the CEC Chapter 13 Panel investigation into the Mexican corn case highlighted that not all knowledge is in the scientific domain. The Panel concluded that the small, indigenous landholders and farmers, known as campesinos, had significant knowledge of both the *in-situ* properties of landraces and the agronomic and social practices of farming, which could have significance for scientists undertaking field studies and for regulators assessing risks of open cultivation. This knowledge does not conform to the scientific norms, but remains real and potentially important in many contexts. Undoubtedly, there are other knowledge systems that may have a significant impact on our understanding of the opportunities and risks of other transformative technologies. Second, although the pre-existing scientific institutions were eventually able to address the scientific question raised by Quist and Chapela, they proved to be too slow to meet the demands of society. At one level, the CEC Chapter 13 Panel process helped to delineate the knowledge claims, but it was only temporarily able to provide a forum for public engagement. More is clearly needed if we want to seriously commit to a post-modern world of knowledge generation and use.

## CONCLUSIONS

Transformative technologies by definition transcend the boundaries of the known. The Committee for the Study of Invention, commissioned by Lemelson, MIT and the US National Science Foundation, concluded that the inventive, knowledge generation process is fundamentally driven by individuals who:

> are not trapped by what they know or think they know. They are boundary transgressors. They mobilize their knowledge flexibly, selectively, and critically. They often abandon what is 'known' in several senses – setting aside previously effective approaches that do not seem helpful in a specific case, bracketing knowledge as not helpful, and challenging prior knowledge as perhaps false or flawed.[54]

In order to predict when, where and how these boundary transgressors will emerge and precipitate change, we must have an understanding of the system(s) they are transgressing. The 'republic of science' remains strong, but is no longer the only authority in defining knowledge. Regardless of one's opinions about the value of post-modernist public engagement, it is here to stay and has precipitated a wide range of appeals to authority beyond the peer-based structures of normal science, including reference to formally sanctioned intergovernmental structures (such as the OECD, FAO and the US National Academy of Science) and to newly sanctioned or purpose-built venues (such as the Nuffield Council on Bioethics, Royal Society of Canada expert panels, US President's Council of Biotechnology Ethics, CBAC and the CEC). How we balance these conflicting knowledge claims and use the resulting knowledge will in many cases determine the range, scope and scale of innovation.

## NOTES

1. Jasanoff (2004: 4–6).
2. Caufield (2002).
3. Shapin and Schaffer (1985: 25).
4. Polanyi (1962: 60).
5. Merton (1942: 268).
6. Merton (1942: 268–9).
7. Ibid., 273.
8. Ibid., 274.
9. Ibid., 270.

10. Ibid., 272.
11. Ibid., 276.
12. Thornton (2005) 'Karl Popper', *Stanford Encyclopedia of Philosophy*, accessed 26 August 2005 at http://plato.stanford.edu/entries/popper/.
13. Thornton (2005) argues that 'Popper's theory of demarcation is based upon his perception of the logical asymmetry which holds between verification and falsification: it is logically impossible to conclusively verify a universal proposition by reference to experience (as Hume saw clearly), but a single counter-instance conclusively falsifies the corresponding universal law. In a word, an exception, far from "proving" a rule, conclusively refutes it.'
14. Laestadius (2000).
15. Nightingale (1998).
16. Asheim and Gertler (2005).
17. Mokyr (2002a). See also Mokyr (2002b), Chapter 2, for a masterful treatment of the historical evolution of these systems.
18. McNeill (1996) focused on biological nomenclature, involving the scientific naming of animals, plants, fungi, protists, bacteria and viruses.
19. Davis and Heywood (1963).
20. Mokyr (2002b: 29).
21. www.palaeos.com/Systematics/Linnean/Linnean.htm#Linneus.
22. Hawskworth (1995).
23. http://www.iczn.org/new per cent20the per cent20Code.htm.
24. McNeill (1996).
25. Hawksworth (1992).
26. The periodic table contains a great deal of information and people familiar with its structure can quickly determine a significant amount of information about an element. The table arranges the elements left to right and top to bottom in order of increasing atomic number, with the different rows of elements (called periods) signifying the highest energy level an electron in that element occupies in the unexcited state. The number of electrons increases with each of the seven rows. Elements that lie in the same column on the periodic table (there are 18 of these 'groups') have identical valance electron configurations and exhibit similar chemical properties (for example all the group 18 elements are inert gases).
27. Since then a wide range of alternate systems have been developed and adopted in parts of the world. The Universal Decimal Classification system of library classification, developed by two Belgian bibliographers at the end of the nineteenth century, is based on the Dewey decimal classification, but uses auxiliary signs to indicate various special aspects of a subject and relationships between subjects. The Colon classification system, developed by S. R. Ranganathan and widely used in India, is a faceted (or analytico-synthetic) classification where colons separate facets into 42 main classes that are combined with other letters, numbers and marks. The Bliss bibliographic classification, developed by Henry Bliss in the 1940s based on Ranganathan's approach, provides a full facet analytical approach to differentiate literature into 13 categories and various sub-categories; this system is widely used in the UK. Many university and national libraries use the US Library of Congress Classification system, or some derivation of it, to catalogue their collections. The classification, developed by Herbert Putnam in 1897, divides subjects into broad categories and enumerates the nature of the work. Each of these systems is either privately or collectively managed and is marketed to libraries and others on commercial terms.
28. This section draws conceptual and factual information from Wikipedia entries on Mathematics and the History of Mathematics and from ThinkQuest Library on Mathematics History, accessed on line at: http://library.thinkquest.org/22584/
29. The study of quantity involves natural numbers, integers, arithmetic and more recently number theory (the study of properties of whole numbers). The study of structure through abstract and linear algebra involves the investigation of various abstract number systems and vector spaces. The study of space is grounded on geometry (Euclidean trigonometric and non-Euclidean analytic, differential and algebraic geometry) and involves the examination of multidimensional phenomena, such as general relativity and topologies. Investigations of

change engages calculus, with the concepts of functions, differential equations, real analysis, complex analysis, functional analysis and dynamic systems, all relevant to the study of quantum mechanics and dynamic systems, including chaos theory.

30. Boulding (ND: 1–2).
31. http://www.unesco.org/iau/directories/index.html.
32. DeNardis, L. (2001), 'Focus: Shared governance,' *The Presidency*, Fall.
33. Woodley (2001).
34. *The Economist* (2004), 'Access all areas,' 7 August, pp. 64–5.
35. Baty (2005).
36. http://www.biomedcentral.com/.
37. www.pnas.org.
38. www.pubmedcentral.nih.gov.
39. Personal correspondence with the editor of a reasonably well-established open access, social science journal.
40. Goodman *et al.* (1994).
41. Giles, J. (2005).
42. Gibbons *et al.* (1994).
43. Anon (1999), 'Bioengineering leaving a bad taste', *Scientific American* 05.
44. Ewens and Pusztai (1999).
45. Royal Society (2002).
46. UNDP Agenda 21, Chapter 16, accessed 14 August 2006 at http://www.un.org/esa/sustdev/documents/agenda21/english/agenda21chapter16.htm.
47. Losey *et al.* (1999).
48. Powell (1999).
49. Ibid.
50. Quist and Chapela (2001).
51. Ortiz-Garcia *et al.* (2005).
52. CEC (2005).
53. CEC (2004: 16).
54. Committee for the Study of Invention (2004).

# 10. Governing invention

Imagination is more important than knowledge.

**Albert Einstein**

## INTRODUCTION

For better or worse, the process of invention and the nature of inventors is for many of us grounded in the Victorian novel *Frankenstein*, written by Mary Shelley. Her gothic novel portrayed Dr Frankenstein, the inventor, as a driven, reclusive, highly educated but slightly demented scientist. His experimental method – with the goal of seeking to create life, working alone and without peer review and using his hunchbacked manservant to burgle fresh graves for body parts – offends our modern sensibilities about the role of science, proper scientific method and the dominion of mankind. Finally, his invention, the reanimated 'monster' is loosed upon the world and humanity, without forethought, risk assessment, post-release monitoring or, in fact, any controls at all. All in all, Shelley's story triggers our basic fears that inventions are fundamentally risky or uncertain ventures fraught with potential hazards and without proper governance and control.

When genetically modified foods entered the global public consciousness, triggered in 1998–9 by the studies by Drs Pusztai and Losey, it didn't take long for the NGO community to alight on the analogy of Frankenstein and his 'monster' and to dub the resulting foods 'Frankenfoods'. Even the mainline press, including the quality broadsheets and some leading business magazines, enjoyed playing with the analogy, using the term and concept in headlines, articles, cartoons and pictures.

The analogy has stuck in the minds of many. In the case of GM foods, instead of Dr Frankenstein, the inventor is often characterized as a group of guileless scientists working for or in collaboration with a number of large, greedy agrochemical companies. Given that general public awareness of the technology arose only after a few specific GM foods had been released into the environment and were embedded in a variety of foods in the food system, many interpreted that to mean that the research was done in some clandestine way and that there was little if any oversight and control of the inventive process or the invention itself. Furthermore, the startling assertions in the studies by Drs Pusztai and Losey simply convinced many that the technology was like Frankenstein's monster, loosed upon the world and humanity without forethought, any post-release monitoring or any controls at all.

Although perceptions certainly are influential, they seldom are a perfect reflection of reality. This is especially true when the impressions are an amalgam of Victorian fiction, journalistic style and incomplete information. In the case of GM foods, there were extensive systems in place around the world to define the focus, scope and structure of the inventive enterprises. In the first instance, the inventors were far from isolated from outside influences. The individuals engaged in the inventive process were well immersed in their environment. As modern scientists, engineers and plant breeders, most of them were heavily engaged in their professional associations, in their industry and in their communities. They also used an array of research resources – money, germplasm, technologies and laboratories – that were owned, operated and governed by a variety of formal institutions, including granting councils, peer associations, philanthropic foundations and joint-stock companies, which in turn are governed by an array of guidelines, rules, protocols and regulations.

The combination of individual initiative and creativity and institutional support and control has led to a complex web of interlocking and overlapping governing systems that has significant control over who does what, when, where and how. Some of the institutions exercise overt, relational power (that is the power of A to get B to do something they would not otherwise do) whereas others exert softer, structural power (that is the power to shape and determine the structures within which others have to operate). This chapter first inspects the array and nature of the main institutions governing the inventive process and defining 'invention' and then examines how those institutions have engaged in governing many of the biotechnology-based inventions in recent years.

## DEFINING AND ORGANIZING INVENTION

Invention is a highly purposeful activity. Unlike research, which is the search for meaning (commonly called know-why knowledge), invention is the process of 'devising and producing by independent investigation, experimentation and mental activity something which is useful and which was not previously known or existing'.[1] The key is that the outcome of the search must be a useful product, process or technology. In that context, invention is not a clean, esoteric deductive exercise like research, but inevitably involves an ad hoc mixture of deductive and inductive reasoning, deep understanding and context, imagination to see beyond the here and now to a different future, a dose of stamina and a good measure of luck.

Invention is more akin to technology than it is to science. Although there is now a high degree of co-dependence and co-development of science and technology (noted in Chapter 2 and to be discussed further below), this has not always been the case. Joel Mokyr, a professor of economic history from Northwestern University, asserts that before 1700 or so, technology was a 'world that had engineering without mechanics, iron-making without metallurgy, farming without soil science, mining without geology, dye-making without organic chemistry and medical practice without microbiology and immunology'.[2] Michael Polanyi reports that as recently as the great London Exhibition of 1851 only one invention – the Morse telegraph – depended on scientific progress within the previous 50 years.[3] Rather than science leading technology, in many cases technological improvements preceded scientific understanding and frequently precipitated scientific investigations.[4] Although there has been a rapid convergence of science and technology in recent years, not all technology has a scientific underpinning. Electrical and chemical engineering, for example, are critical contributors and drivers in moving science to technology – being necessary for scale up – but are not reducible to applied physics and chemistry because aggregation of the component parts leads to non-linearities.[5] Furthermore, there is a strong argument to be made that although scientific theory may govern parts of the experimentation process (as in theory offers an hypothesis that can be tested), the precise design of an experiment and the translation of the results into a new product or process are seldom driven by theoretical structures.[6]

The use of invention as a measure of change is a socially constructed concept. Although some assert that the underlying concept of a 'culture of improvement' can be traced back to prehistory,[7] the modern concept of invention as a distinct novel, useful and non-obvious change in some product, technology or process got its modern roots in the time of the Renaissance and Enlightenment. The term 'invention' is inextricably linked to the concepts of ownership and change that emerged during that time. The first definition of invention can be traced back to the late-feudal period of the fifteenth century. Though one can clearly see the concept of change defined and codified in the first patent laws promulgated in 1474 in Renaissance Venice,[8] the modern definition used today is more firmly rooted in Enlightenment England. In the first instance, the evolving common law system in England accepted and confirmed that individuals have the right to the fruits of their labor.[9] As the Enlightenment period matured, the British government developed and enlarged the concept through the Royal patent. It was originally used in 1557 to govern the publishing industry, allocating a fixed term of exclusive use-rights to stationers for publications in exchange for their managing the content. Over the succeeding years, the patent was extended to other groups, including private consortia seeking to develop and exploit new colonies, to the guilds to encourage the importation or development of new technologies or crafts and, finally, to any and all 'inventors' of most novel, useful and non-obvious change in products, processes and technologies.

Many people have attempted to improve or amplify the legal definition of invention embedded in patent laws around the world, but the key points remain. Generally, inventions are defined as those technologies, processes and products that are novel, useful and non-obvious. Organizational innovations and business processes are also now generally viewed as inventions but basic knowledge, theories, concepts and ideas don't count as inventions in the context of patents and, generally speaking, are not counted as inventions in the economic and social sense. Although undoubtedly valuable, they are not directly 'useful' and hence may be a precursor to but are not in and of themselves counted as inventions.

The concept of an invention as useful change is both drawn from and governed by the law and its role in defining and adjudicating property rights. As our technological capacity has expanded, so have the rules defining what constitutes an invention. The main driver for change has been the United States. As both the single largest source of and single largest market for new inventions in the past century, the US plays a dominant role in almost all debates about inventions and their uses. Over the years the US has extended and entrenched private property rights to a range of technologies and their uses. In the 1980s, a number of landmark rulings related to patenting living organisms, business practices and computer code were instrumental in defining the nature of a number of transformative technologies. In 1980 Mr Chief Justice Burger, delivering the opinion of the US Supreme Court on *Diamond* v. *Chakrabarty*, concluded that US patent law provides for patenting 'anything under the sun that is made by man'. Over the following years a number of appellate court decisions extended patent rights to multi-cellular plants, animals and to living human cells. Meanwhile, in 1981 the US Supreme Court in *Diamond* v. *Diehr* overturned a decision of the Commissioner of Patents and approved a patent on computer software related to the curing of rubber. Finally, in July 1998, a federal court upheld a patent for a method of calculating the net asset value of mutual funds, effectively extending patents to business practices.[10] Essentially the court ruled that patent laws were intended to protect any product, process or method, whether or not it required the aid of a computer, so long as it produced a 'useful, concrete and tangible result', effectively legitimizing both software patents and methods of doing business, and paving the way for Internet-related patents.[11]

Although this concept of invention (and the corresponding rights and obligations) was initially restricted to the United States (as patents are only national instruments), it has been internationalized through a number of institutions. Many individual countries (such as Canada and the EU) have simply adapted and adopted the US policies to their system. Meanwhile, there has been a multilateral effort coordinated through the World Intellectual Property Organization to extend property rights internationally, through a range of international conventions related to patents, trade secrets, copyright, trade marks and various other property mechanisms. Although those voluntary processes had some effect, the pressure to conform has increased since the 1995 negotiation of the World Trade Organization Agreement and its subsidiary Trade-Related Aspects of Intellectual Property (TRIPS) Agreement. That bundle of agreements requires all member states (149 countries in December 2005, with another 29 countries negotiating to join) to either grant patent protection to the full array of human inventions or, in cases such as plants, other *sui generis* or purpose-built systems of protection. Given that all member countries are now bound by the TRIPS rules (developing countries were given until 2006 to conform), the pressure is now on virtually every country to provide a full array of protections for invention. This both extends the legal rights of inventors but also extends the modern, market concept of invention to many countries where collective and traditional practices either undervalue or spurn individual enterprise and ingenuity.

## INVENTORS AND THE INVENTION PROCESS

In the first instance, the nature and structure of the inventive process is fundamentally a governing system. Who pursues change, why they want it and how they go about structuring their efforts all delimit the nature and scope of the inventive effort. Over the past five centuries there have been a range of individual and collective efforts.

Invention starts with people. At any time, only a few of us have the predilection to be an inventor. One way to classify the population is to distinguish them based on their willingness to engage in using new technology. Although perhaps a majority of the population is risk averse and basically conservative, there are still a large number of potential promoters or agents of change. Among those disposed to view change positively, the least engaged are the adopters. These are people (and institutions) who are willing to purchase and use off-the-shelf systems for routine applications. These individuals are often characterized as having a strong survival instinct but a relatively short-term planning horizon. Sometimes this is because of the character of the individual; at other times the risk aversion is due to external constraints, such as insufficient capital or vulnerability to other unmanageable risks. Another segment in society is made up of adaptors, who generally are aware of the balance of threats and opportunities and have the capability and capacity to take a nascent product or technology and make it work in their own circumstances. These individuals are often early users and key partners in making a new technology succeed and thrive. Nevertheless, adopters and adaptors are not fundamentally inventors. In the context of the earlier discussion about types of technological change, they are induced innovators.

The quintessential inventor, be it Dr Frankenstein, Thomas Edison or Alexander Graham Bell, is a boundary transgressor. The Committee for Study of Invention concludes that the inventive, knowledge generation process is fundamentally driven by individuals who:

> are not trapped by what they know or think they know. They are boundary transgressors. They mobilize their knowledge flexibly, selectively, and critically. They often abandon what is 'known' in several senses – setting aside previously effective approaches that do not seem helpful in a specific case, bracketing knowledge as not helpful, and challenging prior knowledge as perhaps false or flawed.[12]

They ask questions never asked and ponder answers beyond the here and now. Various studies tend to conclude that inventors are qualitatively different from the majority. Individuals who end up as inventors (as few begin with that goal in mind) tend to be committed to practical action, are nonconformists, are passionate about their work, are unquenchably optimistic, exhibit a high tolerance for complexity and ambiguity, are willing to delay gratification and often are highly critical about their own work. They also seem able to embrace failure and are willing to abandon knowledge that is too constraining.[13] These individuals are keenly alert to problems and opportunities drawn from a deep knowledge of their area. They have wide interests and are curious about the world, asking questions like 'How does it work?' and 'How could it work better?' In a practical sense, inventors tend to be non-linear, eclectic thinkers. They tend to generate many possible solutions to a problem, from which they will choose the best approach(es). They also have the ability to generate tunnel vision, pursuing a problem and its solutions with blinkered determination. Although some of these characteristics can be applied to other people in society, such as research scientists, public figures and humanists, the key feature that distinguishes them is that they are driven to solve problems. The payoff for inventors is seeing their ideas used.

The modern inventive process is built on the ingenuity, imagination and determination of individual inventors but the relationships of those individuals with others is a key factor in determining the nature of the system. Some inventors operate independently of others while some are part of groups in an ever widening array of institutional settings. Although there may have been a desire and predilection to tinker and improve our material world for eons, the modern inventive period began in the early Renaissance period, when princes and kings personally encouraged and sponsored their greatest philosophers and mathematicians to direct their efforts toward matters of material interest to the state – often focused on the improvement of the crafts or the advancement of military weaponry. The life and times of Galileo and Leonardo da Vinci, among others, followed this path. These Renaissance scholars accepted patronage of one or a succession of rulers in exchange for working on solving problems and inventing technologies related to navigation, transportation, industrial development or war.

As the Enlightenment progressed, the personal role of rulers diminished and the individual inventor came to the fore. Scholars, practitioners and engineers, all largely 'gentlemen' sought the esteem of their peers by solving technical problems and inventing new technologies and products. This system, which was most prevalent in the UK, was institutionalized with the development of various elite networks, such as the Royal Society and various informal groupings of gentlemen seeking to compete and correspond with others with common interests. Although this system led to some significant breakthroughs, the resulting research efforts were for the most part a bit hit and miss and not exploited to their fullest. It was generally assumed that the esteem was as important, or at times more important, than any financial rewards.

As the industrial economy expanded in the early nineteenth century and some of the earliest inventions were commercialized and generated vast new wealth, it became clear that invention could be a lucrative activity. The corresponding rise of the industrial and commercial class in the UK and elsewhere brought a new breed of men into the field, men who were less bound by class and culture than their earlier British counterparts. These men were more willing to get their hands dirty and to husband their inventions into use. These two factors converged, bringing forth a new group of inventors driven more by the money than the glory.

By the 1870s, a synthesis of the two apposite models had emerged. A new group of visionary scientists and inventors applied themselves in a concerted and sustained way in what some have described as 'invention factories'.[14] These factories, exemplified by the labs operated by Thomas Edison, Elmer Sperry and Hirum Maxim, involved leading scientists or engineers who were in charge of envisioning a problem, assembling the capital and leading the inventive effort; in support of their effort, these factories employed an array of technicians, scientists and engineers who were tasked with realizing the visions of the leader.[15] These labs produced hundreds of inventions often spanning an impressive range of technologies. Edison took out more than a thousand patents in his name, Sperry more than 350, and Maxim's numbered in the hundreds.[16] Although fame and fortune were part of the mix, some argue that leading inventors like Edison were primarily motivated by the unique opportunity to change the world. Philosopher Alfred North Whitehead has asserted that 'the greatest invention of the nineteenth century was the invention of the method of invention'.

These invention factories dominated until almost 1920, when the era of industrial labs emerged. The idea of formal corporate R&D groups was the brainchild of a few German chemical companies in the late nineteenth century and gradually was adapted and adopted by major US companies.[17] By the 1920s most of the big US companies – including Dupont, Eastman Kodak, General Motors, Ford Motors, AT&T and RCA – had in-house R&D units and by 1931, for the first time, the US Patent and Trademark Office assigned more patents to corporations than individuals.[18] Since then, corporations have played an increasingly dominant role in the inventive process.

The unique experience of the Second World War, when entire economies and societies in a large number of countries were totally mobilized and devoted to war efforts, created the foundation for a new innovation system after the war. The experience of publicly directed research consortia involving groups of private and university scientists and engineers (such as in the Enigma and Manhattan projects) illustrated the power of new and larger collaborations. Vannevar Bush, director of the Office of Scientific Research and Development which oversaw the Manhattan Project during the war, in 1945 issued a manifesto for this new world. Drawing on his experiences, Bush recommended to the President that the relatively ad hoc collaboration of industry, government and the academia that emerged during the war should be formalized through the creation of an ongoing institution to fund and govern these initiatives. In 1950 the National Science Foundation was finally created, solidifying the relationships that emerged during the war. Since then, the NSF and various other US federal government actors have either directly undertaken or funded research. This move to directly fund or undertake in-house research spread to other OECD countries, with most of them operating extensive peer-reviewed granting councils, extensive intra-mural government research and a wide array of special purpose collaborative structures to engage industry and academics in further research. The next major change in the system came in 1981, with the implementation of the Bayh-Dole Act,[19] which transferred ownership of any university-based inventions resulting from US government support to the universities and required them to undertake efforts to commercialize those inventions; in the past, the results of federally funded research generally were owned by the federal government. The implications of this were far-reaching, with university researchers and university administrators now taking a much greater interest in undertaking inventive activities.

Since then, most national and global research systems have become far more complex. The system has evolved and mutated, with an array of public, private, philanthropic and university-based research systems now involving a multitude of actors, including national and regional state-run research facilities, private sector labs, university labs, a few remaining invention factories, philanthropic or public interest research groups, international scientific research centers (such as the 15 CGIAR centers) and an amazing variety of independent inventors working in their basements and garages. Furthermore, a large number of practitioners continue to tinker in the field and in the factories, finding new ways of doing things better. In addition to the wide array of centers of inventive effort, a dizzying assortment of new proponents, promoters and funding partners has emerged and engaged in the process, ranging from regional, national and international governmental organizations, all sizes and scope of private ventures and a large number of civil authorities – including values-driven NGOs (such as the David Suzuki Foundation), a range of industrial collectives (such as the DRAM chips consortium and various farmer-financed breeding programs) and a few epistemic communities (such as The Royal Society). There is a view that a new style of invention has surfaced, epitomized by the closely knit, highly competitive information and computers complex in Silicon Valley, California. This system exhibits attributes highlighted by Gibbon's mode-2 knowledge, Etzkowitz and Leydesdorff's triple helix system of industry, government and university alliances and Klein and Rosenberg's chain link model.

In the rest of this chapter, I examine the various roles undertaken by the governments, the private sector and various civil authorities, and then put their respective efforts into context with the evolving regional, industrial, sectoral and global innovation networks that we can see in various parts of the world.

## EFFORTS TO GOVERN INVENTION

At the most basic, 'what gets valued gets done'. The most recent OECD data suggests that the 30 OECD member states – although representing only 18 per cent of the world's population are the source of 59 per cent of the world's GDP, 75 per cent of world trade and more than 95 per cent of the world's patented technology – expended on average 2.24 per cent of their combined GDP on R&D activities, worth an estimated $640 billion in 2003.[20] The impetus behind this massive effort is almost always some articulation of human needs and visions. The critical issue is whose needs and visions get valued and how that value is translated into action. This raises the basic journalistic queries related to who, why, what, where, when and how.

We have reached the point where there are a wide range of actors in the system, each identifying some of the things they value and translating those wishes into commitments of money, materials or contextual knowledge and *ex ante* indications of interest and willingness to buy, use or promote an invention upon introduction.

The reasons why these different actors are engaged varies widely. Economic opportunity (in terms of profit for companies and inventors, jobs, incomes and growth for governments or empowerment and social improvement for various NGOs) drives many actors. Some actors seek to remove technical constraints on our society and economy, in order to re-order the activities or outcomes of our societies. Others seek to avoid what they view as undesirable outcomes, be that a risky impact or an ethically compromised process related to specific inventions. Some activities, be it stem-cell research using human fetal tissue or transgenic transformations that transfer genes between different species and different kingdoms, offend groups; governments and civil authorities often use regulatory authority, financial incentives or penalties, legal action or public opinion to moderate research to their norms.

The answer to the 'why' question often, but not always, determines what gets done, or in some cases what is not done. Many inventions are simply serendipitous outcomes of processes seeking other goals. Nevertheless, why an actor is engaged in promoting or hindering an invention will often determine what gets done. Increasingly it seems that money imparts more impetus, influence and meaning to researchers than do other less tangible but still important goals.

Where and when appear to be linked to the answers to the questions of who, why and what. Much of the activity directed at satisfying those engaged in identifying proactive research is becoming networked, with actors from the three broad domains of the public, private and collective sectors combining at different stages in the research lifecycle. The different stages are frequently driven by different groups, with the earliest stages often led by researchers themselves or in collaboration with NGOs, granting councils or various philanthropic granting organizations, such as the Rockefeller, MacArthur and Gates Foundations. As an invention becomes more tangible, costs and benefits become delineated and private capital can be more easily mobilized through companies (such as Monsanto and Microsoft), venture capital corporations or other financial intermediaries. Each of these actors operates in a different time scale, uses different metrics and decision tools and has preferences for different locations or communities. Ultimately, the balance of power in the research world will determine where and when things will happen.

Recently, many actors in the research world have directed more of their attention to the question of how research is conducted, rather than focusing on the who, why, what, where and when. While in the past there may have been a tendency to accept that the ends of an effort would justify (or at least mitigate) the means, in the current post-modern context ends seldom are viewed as justifying means. Rather, means are now viewed as an integral part of the ends. Hence, there is significant effort throughout the research supply chain, especially focused on the inventive and gestation processes, on establishing an acceptable process and procedure for invention and on proving up that invention. In the context of the inventive step, there is significant debate about how money, material and contextual knowledge are assembled either to directly push the supply of new inventions or to pull new inventions into the market.

## Government

Governments very openly and transparently exert both relational and structural power in an effort to suppress unwanted inventive effort and to nurture desirable outcomes. They use a range of tools, including laws and regulations, direct and indirect funding, selective purchases and moral suasion.

Even before governments engaged in defining invention through intellectual property rights systems, governments ruled by fiat – an 'authoritative pronouncement, decree, command, order'.[21] These fiats could stipulate that someone, or more likely some group, would be prohibited from doing something, or alternatively they would be required to undertake some inventive activity to sustain their position. For example, governments for a long time were highly protective of the arts and crafts of war, and prohibited unauthorized persons from possessing, manufacturing and improving on those deemed to be in the state monopoly. Similarly, many of the early guilds got their franchises on the assumption that they would further the domestic capacity in one or more of the crafts or arts and, if they did not perform, they could be disbanded or penalized for their dereliction. More recently, states have used their power to outlaw certain types of activity when undertaken without authorization. The prohibitions tend to relate to activities that might be prejudicial or subversive to the state order or that offend the cultural or humanist norms of the majority. Although there has been a strong tendency in developed nations to adhere to John Locke's admonition to tolerate almost anything that is not overtly subversive to the state, many of the world's authoritarian states and theocracies have more narrow views of individual liberty and prohibit a wider range of inventive actions.

Perhaps more importantly, governments use intellectual property rights systems to define invention and determine which inventions can have value and use in their domain. The evolution of the Anglo-American model of protection of intellectual property rights has gone a long way to crystallizing the specific value in key subject areas. As these rules have evolved, there has been a creative tension between assigning strong, exclusive use-rights to inventions as an incentive to private investment and to encourage commercialization and the desire to optimize the forthcoming value from extensive use of any invention. All legally mandated intellectual property regimes include some limitations. Most require full disclosure of any invention so that follow-on research can be facilitated and many operate on the basis that an invention must be 'worked' to be sustained. Meanwhile, how the rules are written, especially those related to the scope (what is covered by the patent and what is deemed part of prior art or non-patentable), duration (how long the patent will last and under what terms it may be opened) and the terms for maintenance (the US Patent and Trademark Organization (PTO), for example, levies escalating maintenance fees for patents) will encourage private investors to undertake a hard-nosed look at the commercial prospects for their invention. In many cases, they find they hold a relatively small and wasting asset.

As important as the legal system may be, governments almost everywhere are heavily engaged in funding and undertaking research. On average, approximately 30 per cent of the $640 billion invested in OECD member states in 2003 was funded by governments, with about one-third of that flowing through higher education and the rest being invested either in intra-mural research facilities or transferred to industry or philanthropic groups to undertake the work. Governments have adopted increasingly sophisticated public choice and business investment models for evaluating what research to support and how to optimize their investment from an economic, social and commercial sense. In addition, the public sector is incorporating new risk–benefit frameworks and ethics considerations in their investment programming, with the result that they now are sending out a complex set of messages to those in which they invest. In the immediate post-war period and up into the 1980s the focus was largely on commercial and economic outcomes and goals. Increasingly after 1980, investment decisions were made in the context of considerations about the environmental, health, social and political risks of an investment (to the point that some of the larger research investment programs have been subjected to environmental impact assessment) and to ethics guidelines and reviews. Unlike with environmental issues, where government processes have undertaken the reviews, most of the ethical evaluations have been delegated to agents (such as peer-reviewed granting councils) or devolved to the home institution undertaking the research, where they are required to demonstrate that they have undertaken appropriate research ethics evaluation and monitoring in order to continue to receive the funding.

About two-thirds of the funds OECD governments spend on R&D are channeled through universities. Much of that funding is provided through independent agencies that receive annual block grants from the treasury and then disburse the funds to their target research community. In the US, the National Science Foundation supports fundamental research and education in all the non-medical fields of science and engineering and the NIH makes grants in the health field. In Canada, the job is divided among three councils – the Social Sciences and Humanities Research Council, the Natural Sciences and Engineering Research Council and the Canada Council for Arts. In addition, the Industry department has since the late 1990s created and funded a variety of research foundations with more targeted research goals, including the Canadian Foundation for Innovation, Genome Canada and various multi-year, focused National Centers of Excellence. In the UK, eight research councils offer funding to an array of target areas, including the arts and humanities, biotechnology, engineering and the physical sciences, economics and social research, medical research, the environment and particle physics and astronomy. Similarly, pan-national organizations such as the European Union offer grants under various multi-year, targeted Framework Programmes (the sixth program spans 2002–6 and targets atomic energy, biotechnology, information technology and nanotechnology). Although the models vary somewhat, the basic structures are similar. Government grants, increasingly targeted on a specific subject or method, are assigned to an arm's-length independent agency that initiates a competitive process, which involves peer adjudication and a panoply of checks and balances in the management of successfully granted projects. These agencies act as de facto agents for their governments, ensuring the researchers conform to the overall goals and specific guidelines of their grants.

For the most part, commercial and economic interests remain dominant. The key decision in the US through the Bayh-Dole Act to devolve ownership of intellectual property to universities and researchers directly (and emulated in analogous ways in most other OECD countries) had a major impact on how their devolved dollars influence inventive activity. Now that more people see themselves as engaged in and possible beneficiaries of invention, they are assigning greater value to the inventive portion of their research and to whatever they might invent. Before the Bayh-Dole Act became effective on 1 July 1981 US federal agencies (like most government funding organizations) kept tight control over any intellectual property resulting from research they funded. The Bayh-Dole Act effectively devolved to universities the ownership for all research undertaken with US federal government funds. Essentially, the Act requires any US university using federal funds to have written agreements with faculty and technical staff requiring disclosure and assignment of inventions and specifies that universities are then obligated to disclose each new invention to the federal funding agency and to elect whether to assume responsibility for that invention. If the university chooses to retain title, it must file a patent application and share with the inventor(s) a portion of any revenue received from licensing the invention and it must provide the government with a non-exclusive, non-transferable, irrevocable, paid-up right to practice the invention.

This one piece of legislation turned the university research world on its head. The US-based Association of University Technology Managers (AUTM), representing more than 350 universities, research institutions, teaching hospitals and government agencies in the US and Canada, reports that a survey of 198 US universities, hospitals and research institutes shows that these institutions undertook more than $27.8 billion in federal sponsored research, disclosed almost 16 800 inventions and filed more than 13 750 patents during 2004. They also reported that they had more than 11 000 licenses and options generating more than $1.4 billion in licensing income in 2004 and during the year spun out more than 460 startup companies. Canadian firms reported proportionally smaller but still significant sponsored research, disclosure of patents, licensing and revenues.

Nevertheless, about one-third of all of the funds invested in research and development activities by governments in OECD countries are used by in-house or closely connected research facilities. The US offers a good example of the spread of direct, in-house research capacity. The oldest and single largest program is arguably the one housed in the National Institutes of Health (NIH), which is the primary US agency involved in medical research. Apart from its competitive grants programs, the NIH operates 27 research institutes or centers (each targeted on a specific medical subject), all but two of which are located in or around Bethesda, Maryland. Other programs include the National Aeronautics and Space Administration, established in 1958 to run the space program and now responsible for both long-term civilian and military aerospace research, the United States Department of Energy, which operates 22 national research laboratories spread across the nation, and the Agricultural Research Service, which employs 2000 permanent scientists working on approximately 1050 permanent research projects at more than 100 locations across the country and in five foreign laboratories.

Governments also use their purchasing power in key sectors to influence the nature of inventive effort. Many governments will provide incentives or preferential access to inventions that came from their domestic programs. There has been an effort in recent years through the EU Maastricht Agreement, the North American Free Trade Agreement and the World Trade Organization Agreement on Procurement to constrain government preferences but most governments continue to formally and informally send signals to their inventors through their purchasing.

Finally, many governments lack the political, legal, regulatory or economic capacity to enforce their wishes on researchers – in those circumstances, they often resort to moral suasion to get a desired outcome. Both governing members and opposition members are not above initiating a debate in a legislature to highlight either a positive or, more often, a negative development in a public or private lab. Public censure, with or without any formal sanctioning, often is enough to lead many researchers to simply steer clear of areas of controversy. At other times, governments will set up investigatory panels, review committees or other subjective, value laden processes to highlight priorities and options in an area, in an effort to sharpen the focus of public, private, collective actors or to alert the researchers themselves of public priorities.

## Markets

In the past five years, the private sector in OECD countries has funded and undertaken the lion's share of research, on average funding about 62 per cent annually and undertaking about 68 per cent. The dominance of private actors varies from below 50 per cent in some Eastern European OECD member states to as high as about 75 per cent in Sweden, South Korea, Japan and the US. As noted earlier, the in-house research laboratories are now the dominant source of inventions. Most of that research done in-house is governed under some formal work structure, often reflecting the hierarchical, Taylorist model of management. A range of less structured management systems is used in some firms and sectors. Google and 3M, for example, proudly assert that up to 25 per cent of a research employee's time may be used for work related to the individual's interests and other firms have tried matrix, team or quality circle structures for day-to-day management of workflow in their research facilities.

Much of the industry funded research is important but it seldom leads to transformative technologies.[22] The market system can best pull new technologies into the market by integrating downstream market knowledge with upstream productive capacity. In that sense, established firms are most comfortable dealing with induced inventions where there are readily identifiable market uses and, thus, that can be evaluated in some cost–benefit framework where the decisions can be distilled down to finding projects or initiatives with established business tools. Favorite decision criteria include positive net present values (where both costs and benefits are discounted to the decision point), internal rates of return (where calculated rates are compared with the cost of capital), various real options scenario building (where firms assess the optimal timing of investments) and cash flow analyses (where firms assess the draw and contribution of a project to overall financial requirements of a firm). Each of these analytical tools offers insights into the magnitude of the costs and benefits and nature of the risks a firm may assume in making a choice. These tools tend to be wielded mostly by a firm's central administrative units and adjudicated by teams of senior managers, which puts them under the guidance of corporate law, firm policies, culture and practices and the normal fiduciary responsibilities of those exercising power in a firm.

Although other considerations may come into play, they are usually treated as a secondary concern. On the commercial side, for example, firms may also consider the fit of a new technology with existing product lines, the impact of an invention on unamortized capital assets and the role a new invention might play in securing or expanding market share. The choice of the variables depends a lot on whether the firm adopts a product-based view, resource-based view or knowledge-based view of the firm (as discussed in Chapter 7).

Although these other considerations will have an influence on choosing among competing investment opportunities or at times whether to proceed or not, they seldom will trump a negative net-present value calculation (except where the hubris of a corporate leader or management team overrides the normal project assessment models).

Other non-commercial factors – such as the ethical dimension, the social impact and the distributional implications – are beginning to enter the lexicon of managers and be introduced into management structures of for-profit enterprises. Most multinational companies now have articulated some form of code of corporate social responsibility (Monsanto calls it 'The Pledge'), adhere to some form of CSR reporting (for example, Bayer AG annually issues an online report on its efforts), conform with outside values-based codes (such as as promulgated by the International Christian Chamber of Commerce or the UN Global Compact) or establish their own foundations or trusts to articulate a vision and fund philanthropic works. In many cases firms will adopt a cascading set of commitments across the spectrum of options. Nevertheless, there is significant doubt in the eyes of many social activists about whether these efforts actually have much influence. The prevailing view outside the boardrooms is that non-commercial factors will be paid lip-service and wherever the cost of adhering is minimal they may actually be implemented, but that they are almost never going to be used to trump a 'good business' decision.

Firms are also increasingly influential in the ever changing research networks, consortia and innovation systems that appear to emerge and evolve in many sectors and many regions of the world. Their role and influence in those systems will be discussed in more detail below.

## Civil Authorities

Civil authorities – those that are values-based, interests-driven and epistemically coherent – have an influence that far outweighs their direct expenditures. These institutions on average fund less than 5 per cent of the annual R&D expenditures in OECD countries and directly undertake only about 3 per cent of the actual work. Nevertheless, these institutions have a significant influence on defining what has value and how the inventive process should proceed. Especially for transformative technologies, where the market value of the invention may be poorly delineated, these organizations often either push or pull the technology into existence and then into use.

Values-based groups have been increasingly influential in the articulation of research goals and the identification of appropriate inventions for desired ends. Large, philanthropic endowments established by extremely wealthy individuals and families, one of the traditional values-based actors in advanced industrial economies, have now become almost part of the establishment.[23]

The Carnegie, Ford, Gates, McArthur, Rockefeller and Welcome foundations and trusts have been key players in many of the major transformative changes. Their interventions have often provided the early, seed capital for a major change in our economic or social system. The Carnegie Trust invested heavily in public libraries during the nineteenth and twentieth centuries, Ford invested in community capacity building and education, Gates has targeted his wealth in support of creating and transferring technologies in support of public health in the developing world, McArthur has created a range of five-year fellowships to allow ingenious people to pursue their own research, Rockefeller was instrumental in providing much of the funding for the research in Mexico and the Philippines that brought forward the dwarf, high-yielding wheat and rice varieties that formed the foundation of the Green Revolution in the twentieth century and Welcome was a key investor and leader in the Human Genome Project. More recently, large multinational firms have begun to create their own foundations and trusts as extensions of their corporate social responsibilities. Although these foundations are nominally values-based institutions, they are clearly linked with their initiator's corporate interests.

Other civil society groups have engaged heavily in debate about new technologies over the past half century. Various non-governmental groups seeking environmental, social development and justice goals have become major actors, commanding media attention that is often disproportionate to their size or material resources. As a result, they have been able to become dominant actors in the public forums related to new technology run by industry, government and academia – they are viewed as the voice of the masses. The role of these groups has recently been institutionalized: they are being invited to participate formally in the ethics or advisory panels many firms are establishing; they are gaining formal standing in many national processes (in court cases, environmental reviews and legislative committees) and international organizations (both the WTO and OECD, for example, have formal policies for engaging and providing observer and at times participant status in their deliberations); and universities now frequently invite NGO participation onto their research teams, program evaluations and ethics review panels.

While interests-based groups have frequently been seen in and around debates about invention, they have often been perceived as simply reflections of anti-competitive, rent-seeking oligarchies. Whether or not that view was ever valid, interest groups are currently major forces for integrating efforts and defining the interconnections between disparate bits in what would otherwise be disconnected research systems. Patient advocacy groups have had a long history of engaging in the research process, and now they are formally key actors in defining research goals, setting research schedules and delimiting research methods. More recently, industry has taken a greater interest in organizing to collectively manage those aspects of the research enterprise that transcend individual firms. The biotechnology industry, for instance, has organized a range of collective action groups, including CropLife and BIO, to establish common research protocols dealing with ethics, public engagement and risk management, to present a unified industrial perspective on regulation and legislation and to exchange information about research priorities and roadblocks. Other industrial groups, such as the International Chamber of Commerce, have worked to develop more general guidelines and protocols related to the interface between for-profit firms and governments and society.

Meanwhile, the epistemic groups have been active in defining what is known, what principles, procedures, protocols and processes should govern specific scientific domains and what should, or should not, be researched. For example, the Interacademy Panel (IAP) on International Issues, representing 63 national and international scientific academies from both advanced industrial economies (such as the US, UK and Japan) and developing nations (such as Bangladesh and Senegal and those represented through the African Academy of Sciences), issued a statement on human cloning in September 2003, calling for a ban on reproductive cloning of human beings, but recommending that cloning to obtain embryonic stem cells for both research and therapeutic purposes should be excluded from this ban.

As befits a mix of values, interests and epistemically based organizations, there is no single unifying business or management model espoused by this group of actors. But each of these groups is instrumental in defining what we would view as good research or good laboratory practices and in auditing and evaluating performance against those standards. The latest and perhaps most significant development coming from the civil authorities is the array of guidelines, protocols and standard operating procedures laying out good research, laboratory and inventive practice.

Although many researchers and inventors are quite happy to use their own judgment to guide their inventive processes, research organizations and financiers are seeking greater certainty and conformity of practice. In response, an array of increasingly more precise and prescriptive codes of practice has evolved in firms and research labs. In some cases, the research team simply establishes their own code of practice and standard operating procedures (often without much further consideration of auditing or compliance). Beyond general self-articulated principles and processes, some emerging codes of practice are beginning to gain acceptance. One such code, called 'good research practices' (GRP), is increasingly being adopted and used in laboratories in academic and contract research settings which are producing products that may enter the regulatory regime but are not currently covered by other codes. Regulators are beginning to demand higher levels of transparency in products they are assessing – for example, the EU and US now seek detailed information on the inventive process in their consideration of new drug therapeutics – which is putting pressure on inventing laboratories to establish stronger research protocols and transparency. Because of differing research settings and study types, the standards are applied flexibly. In some settings, a study plan is simply documented in a research notebook as the study unfolds, whereas in other settings a written protocol is established and contracted to by the principal investigator. The UK Medical Research Council (MRC), for instance, has a good research practice publication that lays out general principles, drawing and building upon the seven overriding principles of the Nolan Committee on Standards in Public Life: selflessness, integrity, objectivity, accountability, openness, honesty and leadership.[24] The MRC GRP document identifies steps for planning research, conducting research (standard operating procedures for organization, use, calibration and maintenance of equipment and handling of hazardous processes and materials), recording data (guidelines for gathering, storing and retaining experimental and computer generated data in notebooks and electronic records), reporting results (principles for decisions related to publications, authorship, methods of publication and obligations related to corrections of errors and retraction of published findings) and commercialization. Although most of the practices are laid out in a general way – reflecting the diversity of experimental research – the framework provides a useful checklist of considerations that should underpin a successful and responsible research program.

A somewhat more prescriptive and direct set of guidelines related to 'good laboratory practice' (often styled GLP) establish specific benchmarks for performance in the conduct of research that will enter regulatory regimes. These initially started in the regulatory systems in the US and Europe and have gradually been harmonized and extended through consultations at the OECD, so that they now are fairly standard across the 30 OECD member states.[25] These principles and practices set out managerial concepts covering the organization of test facilities and the conditions under which pre-clinical safety studies are executed. Their purpose is to ensure the generation of high-quality and reliable test data (*in vitro* and *in vivo*) related to the safety of chemicals.

Although these manuals of practice more precisely define and lay out principles, protocols and guidelines, their operation remains somewhat unpredictable. Even when the procedures are undertaken under common guidelines and procedures, widely different results can flow. In Canada, for instance, where since 1998 the three national granting councils have operated under common guidelines (the Tri-Council Policy Statement: Ethical Conduct for Research Involving Human Subjects), the operation of those guidelines has been highly variable. One recent example related to a SSHRC funded survey of the experiences of domestic and international students during the first three years of their university studies that was submitted to research ethics boards (REBs) at four universities in Canada.[26] The package was approved in its entirety at one university, with minor revisions at another, with addition of a legal consent form at a third and rejected at one university until a 'highly detailed and legalistic letter' went with the survey. The response rates at the two universities that required the fewest changes were almost double the rates at the campuses that added legalistic caveats, jeopardizing the nature of the work. Other review processes and operating protocols in other research areas have the potential to influence the research activity and results.

Although the public, private and civil authorities are instrumental in defining the 'who, why and what' of the research enterprise, there is increasing realization that the 'where, when and how' is defined in the context of collaborations and networks of researchers, commonly styled 'innovation systems'.

## INNOVATION SYSTEMS

Invention is a human event – it is driven by humans, for humans and happens in humanly constructed environments. But much of the literature on invention tends to look exclusively at the individual inventor or at the role of the inventor's bilateral relationships with many of the discrete agencies or institutions examined above. Although this may effectively and adequately describe many types of incremental invention, it does not reflect the complexity of the innovation processes related to transformative technologies. There is mounting evidence that many major transformations are generated in the context of some network or community that occupies economic, geographic or epistemic space. For want of a better term, all of this literature is lumped under the rubric of 'innovation systems'.

There is extensive evidence that some form of clustering of activity occurs in parts of many industries and economies. Harvard economist Michael Porter used his framework to examine the nature of agglomeration in the competitive performance of firms in more than 2500 regional clusters around the world.[27] That study, and a variety of other cluster studies indicate that at a minimum there is evidence of uneven distribution of economic activity within sectors and across a range of geographies. Although it is not clearly delineated in the literature, there is also some evidence that there is a life cycle to clusters of activity, with high degrees of agglomeration associated with innovative and high-knowledge sectors or stages in a product or technology life and with declining levels of clustering as the product or technology matures.[28] Similarly, the role of different actors and institutions seems to vary within individual clusters as the product or process matures.

Much of the scholarly work on the economics of location can be traced back to the seminal work by Alfred Marshall, the great Cambridge economist who in the 1890s wrote the book that has largely defined our understanding of neoclassical economics. His presentation and analysis of the Manchester coal-based economy, and its interdependencies provides a sustaining metaphor for understanding modern agglomerations. Although no single agglomeration exhibits only one type of interdependence, clusters can be characterized as primarily one of three types. Many agglomerations are defined in the context of specific products or their supply chains.[29] Firms and actors in these systems are assumed to benefit from extensive forward and backwards linkages, embodied in distinct buyer–seller relationships or flagship firms (often multinational enterprises or national champions). Economists tend to characterize the resulting benefits as economies of scope. A second approach is to define the boundary to the system based on geography[30] where economies of scale are reached, particularly in the context of labor and capital markets. Firms in these communities, often in multiple technology or product markets, benefit from the specific efficiencies of co-location in a region. Both of these systems deliver interdependencies, but they for the most part can be monetized and commercially traded.

A third possibility is that some firms coalesce around a technology or knowledge domain, where scholars, practitioners and visionaries create an innovative milieu that generates spillovers.[31] Marshall argued that in such agglomerations, at some point

> mysteries of the trade become no mysteries; but are as it were in the air, and children learn many of them unconsciously. Good work is rightly appreciated, inventions and improvements in machinery, in processes and the general organization of the business have their merits promptly discussed: if one man starts a new idea, it is taken up by others and combined with suggestions of their own; and thus it becomes the source of further new ideas.[32]

These interdependencies are for the most part not traded commercially. Rather, non-transactional systems govern them. Modern economists suggest that this process delivers at least localized increasing returns to scale, which can at least temporarily intensify activity and draw in resources from other activities.

Each of these different models of agglomeration has a set of institutions which contribute to the development and diffusion of new technologies. These interconnected institutions create, store and transfer the knowledge, skills and artifacts which define new technologies.[33] Both the product and geographic types of clusters or agglomerations undoubtedly exist, but neither is likely to be home for major transformative change. Firms in these clusters will often be highly competitive and see most of their activities as zero-sum exercises. Both these systems would tend to bring forward incremental, cost-minimizing and profit-maximizing inventions; there is little on which to ground transformative change in these communities or networks.

Two discrete, independent systems of innovation could lead to mysteries being in the air.[34] One, typified by the computer software industry is based on flexible networks of small and medium-sized firms, often co-located in distinct industrial districts (such as Silicon Valley), and coordinated by a range of commercial venture capital corporations and angel investors. Firms in these communities tend to exhibit both significant volatility and rapid growth. The other type of system, which perhaps better reflects systems in place in the biotechnology world, is based on the universities, public research laboratories, and large firms performing and commercializing R&D. Regardless of which model prevails, no institution can be or is self-contained in its technological activities.[35] All firms, large or small, have to rely on knowledge from other sources. Systems that support a firm's ability to access, absorb and use external knowledge can be critical to the growth of firms, sectors and regions, especially during the early stages of the development of a technology or whenever a technology has a rapidly changing knowledge base, as is typical with transformative technologies.

Although most of the investigations of innovation systems have focused on examining the components of the system (through an institutional review) or outputs (using share analysis of value added or employment), increasingly there is interest in examining what happens inside the black box of the system. This approach has tended to start with a range of increasingly sophisticated surveys of the actors in and outside a system. These survey results have been then used to correlate business practices against outcomes, to determine if and how connections between actors enhance or impede invention and innovation. This approach has yielded a number of observations about the various roles actors may play.

The untraded interdependencies in these systems are largely governed by two mutually supporting types of actors: centrally located institutions and leaders. In many instances, centrally located institutional actors provide a platform for exchange. Central actors could be public research laboratories, formal or informal research collaborations or industry associations. The price of admission and rules of engagement will vary widely, but ultimately these structures thrive and add value if they are able to generate an environment where participants can feel comfortable exchanging knowledge about what they know and what works (or sometimes more importantly, what they don't know or what doesn't work). Many of these systems are formalized in the form of some formal contract or partnership, which makes transparent the rules and enforces the agreements. A number of models have been tried. Australia has Cooperative Research Consortia and the Commonwealth Science and Industrial Research Organization (CSIRO), Canada has created National Centers of Excellence (NCEs) and National Research Council regional laboratories, the European Union has created an array of Framework collaborations to complement the research facilities operated by the member states, Japan and South Korea have actively encouraged their keiretsu and chaebol to create collaborative ventures and the US has legally sanctioned and supported industry collaborative research and development agreements (CRADAs). Each is designed at least partly to create a platform for exchange. Other systems operate less formally, with collaborative relationships revolving around industry associations or informal research groups. In these cases, the rules are often not formalized, but rather they operate based on an implicit 'golden rule' of do unto others as you would have them do unto you. That type of governing system only works where there is a high likelihood that opportunistic action will not lead to large windfall gains and that the community or network will have a sustained life. Essentially, it operates in the context of a 'repeated game', where there is a non-trivial probability that participants will need to interact on a sustained basis, so that one-off opportunistic action will seldom compensate for being ostracized and the attendant loss of opportunities that would follow.

A second group of actors are the formal and informal leaders in a community. The position of a leader will vary depending on the nature and stage of development of an agglomeration. The leader could be a research star, a venture capitalist, an angel investor, a 'big linker' researcher or an industrial actor. Different agglomerations need different leaders at different times. The difficulty is that there is no conclusive way to identify a leader. Each leader exhibits different idiosyncrasies and capacities, which tends to make most analyses highly subjective and value laden. At one level, we are forced to fall back on the psychological, sociological and economic underpinnings to individual and group action. Some of these individuals will be driven by economic reward: some biotechnology research stars in the 1990s crystallized and capitalized their equity in their reputations through their role in new venture startups.[36] Others seek different returns, in the form of reputation, social standing and various internal benefits.

Regardless of the motivation, it is at least possible to identify the relative position and functions of these individuals using social networks analysis (SNA). SNA offers a tool to make invisible work visible.[37] It tracks how knowledge intensive work is done and is used to assess the complex communication channels within a network. As a tool for analysis, SNA views actors and actions as interdependent units, acknowledges that the 'relational ties' between actors are channels for transfer or flow of resources which can provide opportunities for or constraints on individual action. Social network analysis can help to identify boundary spanners, gatekeepers, knowledge bottlenecks and as well as under- and over-utilized individuals or organizations.

The guiding principle behind social network analysis is concerned with the relationship between agents, nodes and actors and in how such units affect one another.[38] The method enables a researcher to better identify subgroups in a given network such as clusters or cliques or to pinpoint isolates or those agents or nodes that appear to be disconnected from the larger network. Such analyses also enable the characterization of such networks into categories such as core peripheries or emergent groups. A number of measures related to density and centrality in social networks are used to examine communities. The concept of density – the proportion of bilateral ties among actors in a system relative to the number of all possible ties[39] – is often used to measure that degree of untraded interconnections in a community. The concept of centrality refers to the importance of a particular actor and the hierarchical nature of an entire network. In general, centrality measures are used to describe and measure properties of actor location in a community or network.[40]

In terms of centrality, it is not just how many connections an agent or node has, but how centrally located are connected actors or agents. How powerful a node or actor is can be measured as a function of their position within a given network. An actor or agent who is connected to many actors who are themselves connected to many actors, thus multiplying their risk or opportunity, would be viewed as a 'big linker'. These big linkers, who often are the research stars or venture capitalists, provide critical leadership at times of greatest uncertainty, especially related to solving complicated problems that require multidisciplinary inputs.

Although one might think that the theory and analytical methods should lead to a deductive set of experiments that ultimately should explain evolution over time, it is far from clear whether there is one, or even a range, of stable, optimal configurations of innovation systems. Some density is required but more is not always better. One theory being propounded is that while uniform density may assist product, supply chain or geographic agglomerations to bring forward incremental inventions, such a configuration could stifle potential transformative changes. Transformative changes may require what has come to be called 'structural holes', that is areas where there is a gap in the governance system that allow truly revolutionary ideas and approaches to germinate and emerge.[41] Rosenberg, using different terminology, has argued that the first industrial revolution was only possible because the 'institutional negative feedback' of the earlier period was removed.[42] Similarly, although some central actors are needed, it is far from clear who they should be and what types of roles they should play.

Although the focus in many supply chains, geopolitical regions and technology fields is on building creative environments that will produce optimal inventive effort, we lack the tools to understand how to do so. Ingenious, creative people and communities arise almost serendipitously and often outside any formal governance system. Once they emerge, they often are captured by government, industry or civil authorities, either during the inventive stage, or more often in the gestation period, when the ideas are reduced to practice, legal and regulatory compliance is achieved and the package is readied for commercialization.

## NOTES

1. Smith *et al.* (2003: 7).
2. Smith *et al.* (2003: 9).
3. Polanyi, M. (1962).
4. Rosenberg (1994).
5. Rosenberg (1994: 144).
6. Rosenberg (1994: 2).
7. Historian Robert Friedel as cited in Smith *et al.* (2003).
8. Smith *et al.* (2003).
9. Locke (1689).
10. *State Street Bank & Trust Co.* v. *Signal Financial Group, Inc.* 149 F.3d 1368 (Fed. Cir. 1998) cert denied 119 S. Ct. 851 (1999).
11. Novos (2006).
12. Committee for Study of Innovation (2004).
13. Perkins (2004).
14. Mokyr in Smith *et al.* (2003).
15. Hughes (1989).
16. Ibid.
17. Ibid.
18. Smith *et al.* (2003).
19. P.L. 96-517, The Patent and Trademark Law Amendments Act, more commonly known as the Bayh-Dole Act.
20. OECD (2005b).
21. OED Online.
22. Hughes (1989: 53).
23. Rosenberg (1994).
24. MRC (2005) 'Good Research Practice', MRC Ethics Series. Accessed on the World wide Web on 7 February 2006, at: http://www.mrc.ac.uk/pdf-good_research_practice.pdf
25. http://www.oecd.org/document/63/0,2340,en_2649_34381_2346175_1_1_1_1,00.html
26. Charboneau (2005).
27. Porter (1990).
28. Lundvall (1992).
29. For example, Roelandt and den Hertog, 1999; Rugman and D'Cruz, 2000.
30. For example, Porter (1990); Rosenfeld (2002); Lundvall (1992); Cooke (2002); Wolf and Gertler (2004).
31. For example, OECD (1999).
32. Marshall (1890: 225).
33. Metcalfe (1995: 38).
34. Malbera and Orsenigo (1990).
35. Metcalfe (1995).
36. Zucker *et al.* (1998).
37. Mead (2001).
38. See Ryan and Phillips (2005).
39. Knoke and Kuklinski (1982).
40. Wasserman and Faust (1994).
41. Burt (2005).
42. Rosenberg (1994).

# 11. Governing the gestation of an invention

> Success is 10 percent inspiration and 90 percent perspiration.
>
> **Thomas A. Edison**

## INTRODUCTION

Inventions have a long way to go to become socially useful innovations. Although germination of an idea is often portrayed in the literature as analogous to Michelangelo's conception of God animating man,[1] the process of translating those creative sparks into use is fraught with complications. The prolific inventor Thomas Edison may not have had any empirical data to back up his views, but his assertion of the 9 to 1 ratio of perspiration to inspiration gives one a sense of the scale of effort required to move an inspiration into use.

The gestation path from inception to birth is tortuous, for both living organisms and for human inventions. We can examine after the fact any successful reproduction or emergence of new man-made technology, institution or product and identify the steps they took but we are somewhat at a loss in identifying when, why and how some inceptions are terminated and some proceed to gestate. In the world of technology and ideas, we do have some evidence to suggest that only about one in ten formal inventions submitted for protection under a patent ever emerge into the market. What we don't know is how many creative sparks fizzle out earlier in the process. A researcher might annotate the idea in a notebook, an engineer might scrawl a note or drawing on a set of designs or an inventor might doodle an idea on the back of an envelope, but often nothing further is done to reduce the idea to practice and bring it into use. The challenge is to begin to understand the processes that govern which inceptions get through the gestation processes and become candidates to be socially useful technologies, institutions or products.

Although all organisms exhibit some of the attributes of a gestation process, mammals represent perhaps the best metaphor for human innovations in that the gestation process involves the fertilized cell being carried in a womb, where it is being nurtured continually by the maternal host while being protected from and immunized against outside threats. Similarly, new inventions are usually 'carried' by a proponent, being nurtured with the expenditure of money and effort to further delineate and strengthen it while, at the same time, it is being poked, prodded and tested by an array of external agents (such as regulators and financiers). Ultimately, the new organism is delivered into the world. In the mammal world, this process can take as little as 16 days for hamsters or 640 days (or almost two years) for elephants. Even then, most of these offspring are rarely independent – they need days, weeks, months, years and sometimes decades of continuing parental support in order to mature and gain their full strength and independence. Analogously, the gestation of inventions into innovations can have widely variable durations and effects. Some innovations, such as some World Wide Web programs, could be conceived in the morning and operational by nightfall. In those cases, the ratio of perspiration to inspiration is much less. For most transformative innovation, however, much more needs to be done to turn an invention into an innovation.

Gestation involves different things happening at different times. In mammals, we know the stages and steps by which a single fertilized cell makes the progression to a full individual, but we aren't sure exactly why and how the single cell multiplies and then differentiates from one cell into the multiple types of cells – such as bone, muscle, eye, blood, tissue and organ cells – that make up a fully developed being. We have an analogous situation in our understanding of how human inventions emerge from the minds of individuals and gestate into fully functioning innovations. We know the steps, but there is no clear set of blueprints on how innovations emerge.

Historian Robert Friedel suggests that the gestation phase involves 'socializing' inventions.[2] The process of translating inventions into socially useful innovations is a multi-staged effort involving multiple actors working on at least three main components. First, inventors are both pushed and pulled to claim the priority of their claim through the intellectual property rights system. Although the IPR system, in particular patents, crystallizes the concept of invention and often has an influence on what and how research is undertaken, it doesn't formally govern the inventor or others until there is a novel, useful and non-obvious invention. An individual or research organization needs a clearly delineated technology, product or process to make an application – patents are retrospectively granted and not prospectively applied. A closely related process entails inventors structuring the management and ownership of their invention to ensure it has the proper support to successfully emerge into use. Second, there is almost always a need to reduce the invention to practice. Virtually all inventions are not fully operational without some further work. Although the inventive step may be there, putting it into use often involves some form of refinement. Going from concept, to design and finally to operation involves a range of efforts. Most technologies, for instance, need to be scaled up: computer programs need to be debugged for potential interface problems and fitted into bigger suites of software; electronic or engineering equipment needs to be adjusted to integrate with larger systems; plant inventions need to be stabilized and backcrossed into commercially relevant plant varieties; and new technologies, products and production processes need to be scaled to existing manufacturing and marketing systems or new systems need to be constructed. This stage is frequently done by the inventor, their assignee or an agent, in a range of laboratories, greenhouses, farms, pilot plants and on the floor of industrial plants. There is an art to this effort – a great deal of contextual, know-how knowledge is involved in fitting a new technology or product into a system. Finally, most potentially transformative inventions trigger some form of government oversight or review of its safety and efficacy. New products, processes or technologies, by the way they are embedded in existing or new production methods or product, have the potential to introduce new or to change the nature of existing hazards in the system, amplify or change the exposure to any hazards and affect the distribution of these hazards. The state has actively engaged in recent years in assessing the range of risks involved in many of these areas, reserving the right to prohibit or delimit the commercialization and use of any new invention. In short, the right of ownership and control and the authority to prohibit others from using an invention do not automatically imply the owner has the right to actually use an invention. That decision flows from one of a range of increasingly complex and interconnected evaluation systems, which can take up to a decade to complete an assessment and issue a ruling.

The inventive process discussed in Chapter 10, though costly and uncertain, is often just the tip of the iceberg of costs of moving an idea into use. Inventors, governments and firms invest a lot in the three gestation processes, which helps to explain the costs of some of the more complex innovations. Moreover, most inventions never emerge from the gestation process – they either fail to meet the legal and financial needs of the owners, are unable to be reduced to practice and scaled up in a cost-effective and timely way or they are rejected as exceeding the safety and efficacy standards set by governments. This triage of inventions means that all of the costs of testing new candidate technologies, products and processes must be recouped from the few inventions that ultimately emerge. Firms and sectoral associations cite high costs for generating new technologies: Microsoft reputedly spent billions to develop each new software operating system; BIO reports that it costs between \$100 and \$200 million to develop a new transgenic crop variety (such as *Bt* corn); and the pharmaceutical industry estimates on average each new patented pharmaceutical drug costs between \$600 and \$800 million.

No single actor has all the authority or capacity to undertake the steps necessary to gestate a new innovation. Different arrays of actors and institutions mobilize different tools and different governing systems to make decisions in these three largely parallel processes. Ultimately, however, the final authority resides in one of two hands. The state has the ability to block a new technology, product or process by fiat, if the innovation is deemed to be unacceptable based on moral, economic, social or cultural perspectives or if it is assessed to be unsafe or, more often, not safe enough. Similarly, the owner has the right to make a decision of whether to use a new idea or not. Companies will often shelve marginal technologies even though they may generate some social benefits. But most importantly, except in a few carefully delimited circumstances, only the owners and their firms can decide when to use a new idea. Although civil authorities have been developing capacity and seeking authority to engage in the various gestation processes and certainly have gained some influence and input to the three main processes, for the moment in most jurisdictions they do not have any substantive decision-making power.

Ultimately, the gestation system, involving the three main actors in three parallel but increasingly interconnected processes, defines when, where, what, why, how and who will bring forward an innovation.

## IPRs AND FINANCING INNOVATION

The intellectual property rights system, based on the patent structure but including an array of other mechanisms, undertakes two fundamental roles. First it defines ownership. It distinguishes between those things that can be owned ('anything under the sun that is made by man' that is novel, useful and non-obvious) and those things that are not subject to private ownership (mainly theories and discoveries that do not meet the three-fold test of novelty, usefulness and non-obviousness). This creates a hard and fast line that distinguishes something new from something that has gone before. Second, and perhaps more important, patents and other IPRs have become the medium of exchange in the knowledge-based economy. Regulators, financiers, research institutes and firms all use patents to delimit rights and obligations and as the basis for structuring the gestation and commercialization for a new technology, product or process.

Although patents are often the preferred means of protecting intellectual property, in practice there are four classes of mechanisms used by innovators to protect their inventions.[3] First, in addition to patents, there are a variety of other legal mechanisms, including trade secrets, plant breeders' rights, trademarks and copyright, all which grant self-enforcing exclusive use-rights to inventors or creators. Second, regulatory regimes established to protect public health and safety also effectively protect private property, by assigning rights and obligations to the owner and thereby impeding imitators; these rights are almost always policed by the state and infractions can result in criminal sanction. Third, a variety of commercial strategies can be adopted by companies to backstop their rights, including linking their inventions to protected complementary technologies, engineering in environmental or industrial constraints or using private contracts (which can be enforced in the courts). Finally, some inventors have chosen simply to publish their findings, protecting their claims and access and, at the same time, precluding others from patenting or otherwise exclusively appropriating their inventions.

Although an invention can be and often is protected using a mix of IPR mechanisms, patents have become the most visible and valuable tool for inventors in most sectors. The patent system is a complex and evolving system. As noted in Chapter 10, it defines what is a new invention, who owns it, its antecedents and linkages to prior art and the duration of exclusive ownership and use.

Patents do grant key rights, but their impact is significantly delimited. In the first instance, in exchange for the limited-term monopoly right (20 years from the date of filing in most countries), the system requires the inventor to fully disclose the inventive step, such that any person skilled in the art should be able to replicate the invention.[4] Moreover, patents are national rights – there is no such thing as a global patent. Although the WTO TRIPS agreement now requires the 149 member states to have patent systems, and the pre-existing Patent Cooperation Treaty makes it easier for an inventor to file for priority in many countries simultaneously, inventors still must proactively apply for and prosecute their applications in each country in which they want protection. If and when the patent is issued (and sometimes somewhat different patents are issued on the same invention in different countries at different times), firms then have the right to exclude others from using their invention. But that right must be enforced by the owner of the patent – the TRIPs agreement requires member states to have laws and courts that can protect and adjudicate any disputes, but it does not impose any obligation on member states to actively promote or protect property rights. Most patent systems also explicitly have processes that encourage inventions to be used or to be abandoned. There has been a trend in recent years for patent offices to levy escalating maintenance fees for inventors or their assignees to maintain their rights. The logic is that over time many patents deliver diminishing returns so, if the cost of maintaining a patent rises, at some point it will be cost effective for an inventor or assignee to abandon their rights and lay open the patent for general use. Similarly, many patent systems have somewhat vague but still compelling rules that require a patent be worked, either by the owner or under license. If a patent is not being used by the assignee, and the assignee refuses to license the invention under 'reasonable' terms and conditions, then most systems have provisions for a third party to petition the patent commissioner for a compulsory license to be issued to allow them to use the technology.

The first thing a patent does is attribute the right of inventorship and assign the ownership to someone. Who has the right to be called the inventor is clearly defined in the various patent laws. Anyone who contributed in an intellectual fashion to the invention must be listed in the application. Adding people who did not contribute or ignoring individuals who did creates the potential for a challenge to the validity of the patent. The patent similarly defines who has the ownership of the patent – usually either the inventor or the inventor's employer (such as a university, private firm, public lab or government department). Different workplaces will have different rules. Most private firms by law or by the terms of an explicit employment contract own any inventions of their employees. The public domain is less well structured. In the US, the Bayh-Dole Act primarily assigns ownership of inventions generated using federal funds to universities, but has provisions that individual inventors may repatriate their ownership and commercialize their own inventions. Other countries vest ownership in the inventor. Patent legislation in Civil Code countries tends to vest the rights in individuals, and some of those rights are deemed to be in some ways inalienable.[5] Some countries, such as Canada, have a mix, with some universities assigning ownership to inventors and some retaining ownership. These structures influence both the range of owners and the distribution of patents among owners. Although corporations only own 46 per cent of all US patents, they own the preponderance of the most highly cited (and by implication valuable) patents.[6] During 1975–98, 85 per cent of the most highly cited patents were assigned to corporations, 9 per cent to individuals, 4 per cent to universities and 2 per cent to non-profits entities. Who is ultimately the owner often has a major impact on whether and when a patent is commercialized and how it fares.

Given this set of converging but still somewhat different rules around the world, inventors and their assignees have the challenge of protecting and gestating their inventions. Who owns and who shares in the benefits of a patented invention can influence how it is used.[7] Regardless of whether ownership is assigned to the inventor or the employer, most employment contracts and granting agencies specify that the inventor should share in the returns of a patented invention, but often the form of that participation is left open. Partly because the value of any specific patented invention is often difficult to measure, most private sector firms try to sever their incentives to their inventors from the value generated by any specific patented invention. Inventors in those institutions sometimes get a small share of royalties or other returns (less than 10 per cent) but more often are rewarded with one-time cash bonuses, stock options or career advancement. The public sector, by its very nature, has fewer options of sharing the wealth – stock options, for example, are not available and rules for promotion are usually well defined in collective agreements or common practice, such that one-off arrangements are seldom possible.

Most public sector organizations tend to share a larger portion of any royalties with the inventors (one study reports anywhere from 35–50 per cent of net royalties, calculated after the costs of the patent application are recouped). The direct involvement of a public sector inventor in the success of a patented invention creates both opportunities and challenges. Some inventors have a good idea of how their invention might be best used and sometimes have the industrial contacts to make it happen. More often, though, academic researchers are impediments to full use of their ideas, as their expectations exceed realistic possibilities and they hold out for unachievable returns.

Practical, operational patent strategies vary widely.[8] Most private firms engaged in purposive inventive effort have some form of intellectual property strategy which involves an evaluation and decision framework and set of well-defined actors who make decisions. Inventors in private sector settings may be invited to participate, but most firms have managers or boards that evaluate the merits of an invention for its in-house or general applicability. Although many firms will consider the strategic implication of an invention in terms of their freedom to operate and other non-monetary factors, financial criteria tend to dominate. Firms will usually undertake some form of financial review, involving a cost–benefit calculation that generates net present value or internal rates of return, similar to what they would do in the inventive stage. This calculation, which is part art and part science, will be undertaken at a variety of points in the gestation process, using a variety of assumptions, including: the appropriate discount rate or cost of capital; the cost of completing the gestation process; the timing of market entry and any induced competition; the potential market price and market size; and the impact on current operations, including the cannibalization of current markets or the destruction of unamortized capital assets. Given that many of these variables will change over time, the evaluation is often undertaken at frequent intervals to capture the dynamic nature of the investment environment. Firms will often do their analysis in-house, but there is a trend in some sectors to contract with experts to provide insight into the uncertainties or variances in key assumptions. The decision of whether to proceed, pause or terminate the gestation of an invention remains a key decision of the firm. Non-financial factors may have some influence for marginal decisions but there is little evidence that they will trump the financial decision. There are many examples where this decision can be altered at even the last moment. Monsanto, for example, shelved its plans to introduce Roundup Ready wheat in the US and Canada in May 2005, even though it reports that it had invested more than $100 million in the venture and was days away from having completed the gestation process. Its analysis of the cost–benefit had obviously turned negative at some point in the final stages of the process.

The traditional, clear demarcation between curiosity-directed research undertaken in university settings and purposive effort in private research labs has become blurred significantly over the years. The internationalization of the Bayh-Dole provision that publicly funded research in universities (and in many cases public laboratories) should be patented and commercialized has had a major impact. Most universities now have binding intellectual property agreements with their faculty and staff and many public and philanthropic granting agencies require scientists and engineers using their funds to actively seek out, protect and exploit inventions. Virtually all universities and public labs now have offices of technology transfer which assess inventions for their value and determine whether to protect and exploit an invention or to lay open the knowledge through publication or some other process. Although these offices have been gaining in power and authority in university settings, it is far from clear how they operate and what incremental value they add. Their business models are quite variable. They can have a variety of motivations and objectives for patenting intellectual property. For-profit firms see patenting as a way to exploit and benefit financially from an invention, either by licensing it or by developing and using it by themselves, but many public organizations do not see monetary gain as a prime motivator. Rather, they often measure success in terms of local economic development or maintaining freedom to operate for their researchers; financial return is only one of the variables they consider. Given the array of objectives, few public institutions have clearly delineated IP strategies. They will undertake a range of evaluations, ranging from some form of cost–benefit analysis to other less formal contextual assessments. Ultimately, there is some evidence that the inventors in the public domain have a larger say in whether their invention is patented, with all its attendant plusses and minuses.

Most enterprises will file a patent as soon as they think they have delineated all of the potential claims of an invention. Alternatively, some inventors or their assignees may decide to file a provisional patent application, which gives them a one-year grace period to complete the application. This approach, which can cost a little bit more than a direct filing, has two benefits. First, assignees can use the one-year grace period to flesh out the claims for the full filing. Many technologies may have multiple applications or uses that could be included in the claims of a patent, widening the scope of the patent and potentially increasing its strength and value. In many cases companies also choose this route in order to test the market. If a buyer or licensee for the invention is found during the pending period, then the buyer or licensee would then often pay the cost of the patent. If no market is found, often the inventor will abandon the patent application.

The inventor or assignee then needs to make a decision as to whether there is any value in preserving foreign protection, if it is still available. Foreign protection can be preserved in two ways. First, the inventor could pursue a Patent Cooperation Treaty application, which ultimately extends the time to file a patent in other countries, usually to about 30 months after the priority date. The second way is to file applications directly into each country where the inventor or the assignee wishes to obtain protection, which is less expensive than a PCT application only if there are a few relevant markets. This decision is important. If an owner aggressively patents their invention in as many countries as possible, then they will potentially expand the market for the invention while simultaneously limiting the use of that product by other companies in other markets. If an owner chooses to only patent in a few counties, then they will limit their ability to exploit the patent to maximize profits. But this choice comes at a cost.[9] Those in the patent business report that provisional US patents cost between $500 and $10 000, depending on the complexity of the technology and its claims. Developing, filing and undertaking the preliminary examination in the US is estimated to cost $3000 to $15 000, including up to $3000 to $5000 for a PCT application (not including the in-house costs of documenting the invention). Translation for and patent filings in a wide range of other countries could add another $100 000 to $275 000. Once the patent is approved, assignees must continue to pay maintenance fees averaging about $1200 per country. In short, costs associated with a patent are higher the more places that you choose to patent.

The filing of the patent and designation of priority is only the beginning of the process. The assignee or owner of the patent is then responsible for determining the timing and nature of the review. It is their job to assemble the evidence needed to assess the inventive step and its related claims and then to initiate the formal assessment of the filing by patent examiners. The lag between filing and granting of patents has recently averaged about 27 months in the US, 22 months in Japan, 24–32 months in Europe and 44 months in Canada.[10] The large and variable length of the gap is a function of the increasing number and complexity of many patent applications (especially for genetic constructs), resource and time constraints in some patent offices and the commercialization strategies pursued by many firms (which often choose to prosecute patents faster in the US than elsewhere).[11]

Ultimately, a patent may be granted. Public servants in a national patent office will undertake a technical assessment of the structure of the patent filing and evaluate the nature of the claims, judging in the first instance whether the claimed invention is patentable and then determining whether the evidence justifies the claim of an invention. Almost by definition this process is a constrained evaluation. Although the patent assessor has access to all previous patents and extensive peer-reviewed literature, they always have limited knowledge, expertise and time to assess all of the claims fully. There is a general view that patent offices try to be fair to both past and current inventors but they appear to err on the side of granting claims and letting the challenge systems deal with disputes. Most governments acknowledge this is far from perfect. Recently there has been an effort by patent offices, working in collaboration with patent agents and lawyers, to implement modest change in the domestic operation of patent systems in various OECD countries. The US PTO, for example, revised its requirements with respect to the utility criterion in January 2001, now requiring patent claims to identify specific uses. The procedures are also being codified: the *United States Manual of Patent Examining Procedure* (MPEP), for example, provides very detailed guidelines concerning the policies and procedures to be followed by staff in the examination of patent applications. These guidelines are public and available for applicants and other stakeholders to better understand how patent criteria are applied. Other counties and various international conventions are seeking to regularize and standardize the patent systems across countries.

A recent development in the patent business is the introduction of an array of new challenge systems. Historically, only the owner of an infringed technology could challenge a conflicting patent grant. Now, most national systems permit other interested parties to challenge the validity of a patent on the grounds that it either embodies prior art or doesn't work. More recently the TRIPS agreement permits an *ordre public* provision to incorporate non-economic values into the patent system. Japan and some member states of the EU have adopted related measures. On a case-by-case basis, patents can be refused when the commercial exploitation of the invention violates public order or morality. The European Directive 98/44 on the Legal Protection of Biotechnological Inventions, for one, explicitly states that processes to use human embryos for commercial purposes or processes to clone human beings violate *ordre public* and morality. In practice, the *ordre public* provision is usually invoked by a third party in an opposition procedure after the patent has been granted. Others provide a broader range of opportunities to challenge patents, including an opposition procedure which provides a forum for raising challenges, typically in terms of novelty and inventiveness, but also with respect to *ordre public*. Australia, the European Patent Office, France, Germany, India and Japan currently have opposition processes and a recent US report has recommended establishing an opposition procedure in the US patent system. Even in absence of a more open opposition procedure, the US has established the Court of Appeals for the Federal Circuit to increase uniformity in appeal decisions.

Different firms in differing sectors will face a range of opportunities and issues as they deal with the patent system. In the first instance, even though patents may be generally available to most inventions, patents are not equally accessed and used by all sectors. The machinery and equipment sector has tended to be the most intensive user of patents, applying for and maintaining patents as a means of contracting and integrating their inventions with those owned by others. Other sectors use patents less intensively: the chemical industry tends to use fewer patents and more trade secrets than other sectors while computer software is more often copyrighted than patented. Many firms supplement their positions with trademarks and other proprietary protection systems to compensate for the absence of patents. In contrast, the pharmaceutical industry and more recently the biotechnology sectors are heavy users of patents.

One issue facing many firms is that many patented technologies or products require access to licenses to use complementary technologies, which at times can lead to a stalemate as competing owners of the technologies may not be able or willing to enter the needed contracts. This is called the 'freedom to operate' problem. Some sectors seem to have overcome the problem. The microcomputer represents a modern example of a machine that is a composite of hundreds of proprietary, patented or copyrighted inventions. Although there have been a number of patent battles over the years, the fact that so many pieces of often evolving and improving proprietary technology can be assembled and made available in a timely way to consumers at increasingly lower prices is viewed by many as a testament to the value of the patent system.[12] Other sectors have not been so fortunate. The biotechnology industry currently faces what it believes is a major freedom to operate problem. The massive effort to patent genes and gene fragments, combined with a few patents on key technologies needed to manipulate genes (such as on gene sequencing, amplification and transformation systems and on popular growth promoters), and the apparent consolidation of many of these rights in the hands of a small group of global agricultural biotechnology and pharmaceutical enterprises, are viewed by many as a threat to the freedom to operate.

Although there is evidence that some firms use their patents to block competitors, many of the apparent barriers may be more economic than strategic. The cost of searching, negotiating and enforcing IPR licenses is quite expensive – some industry participants put the minimum cost for a commercialization license at up to $50 000 per technology, which is prohibitive for most minor applications of those technologies and hence an effective barrier to entry for smaller users. Many of these disputes that cannot be resolved amicably end up in the courts. The courts can effectively handle most difficulties, given enough time. The pharmaceutical industry, which has been using patents and having disputes for many decades, engages in hundreds of patent infringement cases annually, either waiting for the judicial system to adjudicate or truncating the proceeding by resolving the disputes through some negotiated settlement (such as cross-licensing bundles of patents or entering new business arrangements such as strategic alliances, mergers or acquisitions in order to jointly exploit the technology). The agricultural biotechnology industry does not appear to have found a way to effectively handle its disputes. The big players (including Monsanto, Dow, Bayer and Zeneca) have engaged in a handful of patent infringement cases: a few earlier ones were adjudicated by the courts, often to dissatisfaction of both parties, but more recently aggrieved parties have entered broad cross-licensing agreements to resolve their disputes (such as Monsanto-Aventis and Monstanto-Dow). Somewhat surprisingly, there have been fewer infringement cases. One explanation is that, given the limited profits in the highly competitive agri-food sector, the anticipated gains from winning a dispute are too small to justify the high cost of litigating (one estimate is that even the most simple patent infringement case could cost each party in excess of $1 million in legal and consulting fees).[13]

Although there have been calls for the government to intervene in the market, using antitrust provisions, compulsory licensing or some other set of mechanisms to make proprietary technology available to competitive firms, so far governments have remained on the sidelines and left owners to fend for themselves. Some argue that as the agricultural biotechnology market matures, the industry will develop norms that will make most of the patent issues decline in importance.

While a patent application is being evaluated and assessed, most owners are busy trying to assemble the appropriate vehicle for commercializing the technology, product or process. Given that a plurality of all patents and the clear majority of quality patents are in the hands of existing corporations, one would expect that many of the new inventions would be gestated in-house and released as part of a firm's portfolio of products or services. Although economists model the decision systems in for-profit firms as mirroring arm's-length transactions, with shadow prices reflecting costs and benefits of each transfer of resources or materials, the reality is quite different. In spite of more than 200 years of evolving cost accounting and management theory, we still have little in the way of consistent theory and practice. Suffice it to say that an *ad hoc* array of processes, procedures and actors pilot an invention through the gestation process in commercial firms. One observation that many scholars have made is that category-busting potentially transformative inventions tend not to be commercialized first by large, existing oligopolistic or monopolistic firms – they appear to be more concerned at fully amortizing their inventions and maintaining existing markets than changing the nature of those markets.[14]

Transformative inventions have for the most part come from new start-up firms. In many cases inventors or their assignees establish new, purpose-built corporate structures to take an invention to a market. These firms are more likely to consider commercializing transformative inventions because any subsequent market disequilibrium and stranded capital will not be their problem – these factors often won't even enter their calculations. These new ventures almost always need outside investment. In the earliest stages of development they often rely on friends and family as investors, but as the scale of their financial needs rises, they successively tap public development funds, angel investors and finally end up in the waiting rooms of some type of a venture fund. A wide range of financial intermediaries invest in start-up firms. Some will simply be the vehicle for a single sophisticated investor to engage in the market. Others involve pooled funds administered by banks, brokerages, insurance companies, mutual societies, trade unions, cooperatives, government development agencies or other private capital groups. Depending on their owners, they may have slightly different investment objectives (local development funds such as cooperatives and government agencies tend to invest locally), different investment criteria (some will invest only in specific sectors, technologies or market), different assessment processes (some will use highly formalized evaluation models whereas others will make more intuitive decisions) and different modes of interaction with the firms they invest in (some are passive investors and some engage in the strategic or tactical management of the company).

Venture capital corporations have been a recent US innovation instrumental in the entrepreneurial efforts in a number of key sectors. These funds offer a unique mix of services, ranging from access to large, patient capital pools, sectoral and management expertise and a willingness and desire to engage in hands-on direction of their investments. Although ostensibly financial intermediaries, what makes these firms unique is that they often will have a portfolio of a relatively small number of start-up or growth firms clustered in a narrowly defined technology, product or market area in which they invest and aggressively nurture. Some assert that the information revolution, in particular Silicon Valley, would not have emerged without the venture capital industry[15] and others attribute the presence in the US of 1300–1500 new biotechnology firms (sometimes called NBFs) to the presence of the vibrant, aggressive venture capital market.[16] As of late 2005, there were more than 460 venture corporations in the US. Although VCCs' investments consistently equal less than 1 per cent of GDP in every country measured, they are reported to have disproportionately large impacts on new, research intensive firms. It is estimated that more than 70 per cent of all venture funds are invested in US companies and that more than 40 per cent of all companies entering the market through initial public offerings (and a larger percentage of biotechnology and new technology firms) are financed by VCCs.[17] Other markets have relatively limited access to this support.

Other owners – often public labs and universities – have no aptitude, interest or intention of commercializing their inventions themselves and will use the gestation period to find someone who will license or buy their patented inventions and take them to the market. In the parlance of the investment community, these 'one trick ponies' have little chance of emerging from the gestation phase alone – investors will only consider them if they are pooled with other complementary or substitute ventures, such that economies of scale can be realized and risks and rewards can be pooled. While all of this effort is underway, others will be working to figure out how to put the invention into practice and to evaluate the safety and efficacy of the invention in its intended use.

## REDUCING INVENTIONS TO PRACTICE AND SCALE-UP

Few inventions are ready to go into production without further refinement. Just as fertilized cells need to multiply and differentiate, so do new inventive ideas. Most new technologies, products or processes need targeted effort and investment to adapt them to the existing production and marketing system or to prepare them to become the base for a totally new product category. Although transformative technologies always require more work, fortunately they usually can draw upon the technical expertise in existing production and marketing systems.

Generally speaking, reducing an invention to practice and scale-up is done through non-market systems and structures embedded in a firm or through contractual work undertaken in an industrial cluster. Governments in their role as regulators tend to stand back and leave the job to others. The main regulatory directive is that firms can do whatever they wish, as long as any new product, technology or process is not released into the environment or distributed to consumers before it is evaluated and approved by the appropriate regulator. As long as the invention remains contained, firms can pretty much do whatever they want. In some cases, such as in the computer and machinery and equipment field, the government has no specific evaluation process, and so a firm is unconstrained as they seek to figure out how to best introduce their invention. Nevertheless, firms are seldom able to do all the tasks using their own resources. They often will develop teams of managers and a mix of in-house and contracted researchers, engineers and other experts with technical and contextual knowledge of manufacturing, logistics and the market. They will also draw on technical expertise and contextual knowledge that is located in other professional service firms (such as professional engineering and market analysis firms), in industry or consumers' associations (such as farmer commodity groups and patient advocacy organizations), in technical agencies or firms (such as seed breeders at university and public breeding programs or analytical capacity in private labs) and, increasingly, context from non-government organizations (some firms invite selected NGOs to participate in a social evaluation of new products). Each of these has the ability to influence the nature of what emerges from the process. Some of these relationships will be undertaken through arm's-length transactions; others will involve various non-transactional collaborative approaches.

There is significant evidence that the contextual knowledge and technical expertise needed to successfully gestate an invention clusters in locales around major research centers. These centers have the potential to offer benefits of scope, scale and positive externalities. Many of the requisite services, though critical to success, are not needed frequently enough in any individual firm to justify developing in-house capacity. If a variety of firms pool their demand, either formally or simply through co-location, they often can secure an appropriate supply of forward and backward technical supports – that is they generate economies of scope. Similarly, firms in a cluster have the potential to generate significant economies of scale in the labor and capital markets related to the scale-up of new technology. Finally, the co-location of a significant number of competing and collaborating firms, all working on related technologies and products, has the potential to generate significant knowledge spillovers, especially of the contextual, tacit know-how and know-who knowledge that accumulates and is conserved in communities of practice. It has often been asserted that clusters or regional systems of innovation may be even more important for the gestation stage than for the research and inventive stage.

Reducing an invention to practice involves taking the basic recipe from the invention and testing it in applied settings to determine where it works and where it needs to be adapted to work. There is no simple formula for doing this. It is heavily contextual, drawing on the expertise and experience of individuals who usually are part of a community of practice. Many research teams have found that there are modifications and adjustments needed to improve the response rate of many technologies. Although some of this work might be a necessary input to the environmental and human health safety evaluations (as discussed below, the structure and performance of that work will often be highly regulated by government agencies), much of this work is simply to figure out where the technology or product has the best impact. Just because a technology might work doesn't mean that it is worth developing. For example, in the chemical or drug industry, once a new compound has been isolated, purified or constructed, it is still necessary to determine how to regularize the production method, where it might have applications, what market segments might value it most highly and how it might interact with other elements in a production or marketing system. In the computer industry, this might involve debugging and testing how an algorithm works with existing data sets or in the context of various operating systems and special programs. In the agricultural biotechnology industry the ability to transform a plant or animal and integrate it into the global agri-food supply chain is part science and part art.

Scale-up, the second technical step, directly follows from reducing an invention to practice. It involves getting the production and marketing systems aligned to deliver a product or service to users in an acceptable and competitive way. Few products are true monopolies – most new ideas will have to compete with substitute goods or services. Hence, putting the product in its best frame is vital. This involves an array of technical modifications in the production system. This would start with determining and developing the optimal delivery system: backcrossing new traits into commercial varieties; choosing the optimal delivery system and formulation for new pharmaceutical compounds; packaging new computer programs in the context of differing operating systems; or structuring a business model and service package to deliver a new genetic test. Next, technology and product specifications need to be adapted to fit existing production and marketing systems or new systems need to be developed. It is a toss-up which is more difficult. There is little one could say about the changes that might be needed, except that firms will be constrained in what they do by the need to conform to existing standards and protocols in their sector or market, to meet the purchasing order specifications coming from wholesale and retail firms and, ultimately, to satisfy the preferences of end-users. In many cases others will have developed identity preservation, segregation or traceability specifications that will need to be met to gain access to the market; successful products will have these standards and protocols hard-wired into their systems.

The technical hurdles vary widely by industry. In some areas, such as the machinery and equipment and chemical industries, engineers and standardizing bodies tend to dominate the debate. In most of those cases, the supply chains are closely controlled and consumers seldom directly consume the product. This narrows the range of issues to be considered in scale-up. In contrast, inventions arising from the pharmaceutical and agricultural biotechnology sectors tend to involve a larger group of actors. Their products are likely to be consumed widely (by a range of consumers with multiple circumstances) or introduced into a supply chain with a large number of production facilities (onto thousands or millions of farms), so they require a much greater range of skills and expertise to successfully complete the process. In those cases, a range of medical, biological and environmental sciences are engaged, as are a wider range of stakeholders and intermediate and final users. These processes inherently take longer and are more complex. Computer programming is a bit of an intermediate category, as new programs are often debugged by informal networks of users (sometimes called β-testers) immediately before or sometimes upon commercial release.

Ultimately, owners of inventions seek to have reduced their invention to practice and resolved the scale-up problems at about the same time as the intellectual property and organizational structures are in place and the regulatory evaluation of the safety and efficacy of the technology product or process is completed. Although theory would suggest a simultaneous signal from all three processes would be optimal, there is evidence that the regulatory systems often lag the commercial and technical phases.

## RISK ANALYSIS

The third main element of the gestation system is the private and public regulatory assessment of the efficacy and risks related to a new technology or product. This process has changed dramatically in the past half-century. As our understanding of the underpinnings to our technologies and products has advanced and our ability to detect and measure smaller units of matter has progressed, governments are much more involved in examining new technologies, products and processes for their adverse effects and, in some instances, for their positive impacts. Some have said we now have a 'risk society' that is focused on the identification, measurement, assessment and debate of countless risks in our daily lives.[18]

This process is at root an assessment of the effect of a specific, defined product on human health and the environment. For the most part technologies are not assessed; rather the products made with those technologies are assessed. The logic is that although we may use technology in the production of goods or services, it is the explicit configuration of a technology with a specific production or consumption system that will generate benefits or harms. Hence nuclear physics is not evaluated; rather, nuclear physics in the context of generating power or in the practice of medicine is assessed. Similarly, biotechnology is not assessed as a technology; rather its role in generating new or modified microbes, plants and animals or in generating new therapeutic drugs is evaluated. Nevertheless, the presence of a new technology, even when the resulting end product or service is for all intents and purposes unchanged, may trigger an assessment. Thus, technologies can trigger regulatory oversight but the nature of that assessment and the resulting management or communications about risks or benefits is most often related to an explicit application of the technology. Although the advent of biotechnology in the food system has led many politicians, citizens and civil authorities to call for a new way of looking at risks related to transformative technologies, no systems have yet been modified to undertake that task. Most thoughtful observers have simply concluded that the array of potential impacts (both potentially beneficial and harmful) are simply too nebulous to enable us to undertake a technology-based review. So, for better or worse, the discussion about regulating risks of new technologies remains narrowly focused on the products or services produced with the technologies and not on the technologies themselves.

Different products are evaluated with differing degrees of concern. In those instances when the risks tend to be isolated to the individual users in an industrial supply chain or to a consumer's material goods (such that there are no externalities and no considerations of public health or safety), any review is usually left entirely to the private sector and individual consumers. Governments will only play a role if any resulting technology or product is unfairly represented, mislabeled or fraudulently promoted or if it breaches any implicit or explicit contractual obligation (usually generating a civil tort which, although privately prosecuted, would be adjudicated by the courts). Depending on the strength of local tort laws, the threat of civil (often class) action could be a significant incentive for firms to be very careful with their technologies and products. Computer programs, for example, are not assessed by any agency of government before they are released. Rather, the developer will assess the potential risks of a new program before release and will attempt to communicate those risks to users. In some instances, third parties (such as computer magazines, consultancies and other interested parties) may evaluate a program upon release and publicize any strengths or shortcomings. But no one except the owner of the technology is in charge.

In other circumstances, the state has been integrally involved for centuries. Food safety, for example, has been a concern of governments from the earliest times: Egyptian scrolls identify the labeling to be applied to certain foods; in ancient Athens, beer and wines were inspected for purity and soundness; the Romans had a state-run food control system to protect consumers from bad produce; and during the Middle Ages, individual European countries passed laws concerning the quality and safety of eggs, sausages, cheese, beer, wine and bread, some of which still exist today.[19] The second half of the nineteenth century saw the beginnings of the modern food safety system. As food chemistry developed, it provided the capacity to evaluate the 'purity' of a food and to detect the use of hazardous chemicals. In the Austro-Hungarian Empire between 1897 and 1911, for example, a collection of standards and product descriptions for a wide variety of foods was developed as the *Codex Alimentarius Austriacus*. Although lacking legal force, it was used as a reference by the courts to determine standards for specific foods.[20]

The risk analysis framework currently used in most countries evolved after the Second World War in response to a succession of new technologies that posed concerns related to human health and safety and new evidence that the environment could be and was being stressed by industrial activities. The modern pharmacopeias of chemically constructed therapeutic drugs and nuclear power, for example, generated concerns about the environment: Rachel Carson's 1964 book *Silent Spring* focused attention, particularly in the US, on the potential for catastrophe. In response, the US regulatory regimes – particularly the Food and Drug Administration, the Environmental Protection

Agency, the Department of Energy and the Department of Agriculture – expanded the scope of their oversight and began to undertake systematic, pre-market evaluations of a larger array of new technologies and products. This system became codified in various structures as it evolved. In 1983 the US National Academy of Sciences produced the 'Redbook' which laid out a 'risk analysis framework' as a systematic way of dealing with technological innovations in a stable, certain and predictable manner according to scientific principles.[21] Its main innovation was to incorporate science into public policy decision making, thereby importing from the scientific epistemic community a common methodology and a set of rules for dealing with scientific disputes.[22] This insertion of scientific 'rationality' and its related certainty and predictability helped make the system an attractive model for governments everywhere.

The risk analysis framework operating in most developed countries evaluates risk through a set of scientifically based processes that combine the identification and characterization of hazards with assessments of exposure to those hazards. The framework is most commonly divided into three distinct but interrelated functions: risk assessment; risk management; and risk communication. For the most part, the assessment part of the framework attempts to objectively assess the probabilistic outcomes of discrete adverse events, generally abstracting from issues related to risk management or risk communications. The practice is that governments establish a risk threshold for products or classes of products that would reject new products with unacceptable risks but would allow those with acceptable impacts to enter the market.

Risk assessment is defined by the Food and Agriculture Organization (FAO) of the United Nations as 'a scientifically based process consisting of the following steps: (i) hazard identification; (ii) hazard characterization; (iii) exposure assessment; and (iv) risk characterization'. In this model, hazard identification is the determination of whether a particular element of a product is or is not causally linked to particular health or environmental effects. As such, this stage often involves epidemiological and surveillance data, challenge testing and scientific studies. Those macro results often are scaled to sub-populations in society or particular eco-zones. Exposure assessment, sometimes called dose-response assessment, involves determining the relationship between the magnitude of exposure and the probability of occurrence of the health or environmental effects in question. Generally this involves varying the level of exposure until an adverse event is triggered, thereby determining both the causal link and the triggering point for the harm. This often involves examining specific subgroups of consumers that might be most at risk (especially the old, young and immunosuppressed) or the tracking of sentinel species. Exposure assessment, in contrast, is the determination of the extent of human exposure before or after application of regulatory controls.

This includes a description of the pathways through which a technology or product is introduced and distributed in production, distribution or consumption. In short, it involves assigning a probability to the event based on extensive situational analysis of how the system operates and how it would relate to the new technology or product. Finally, risk characterization entails describing the nature and often the magnitude of human or environmental risk, including aspects of uncertainty. This is the final stage of the analysis, where the hazard, exposure and variability of the results are combined to estimate the potential risk of a new or transformed product.

At one level the system appears to be highly objective and deterministic. But as one looks into how the system is created and managed, the normative elements begin to appear. Both normal and emerging science are incorporated in the risk analysis framework through the political process. Scientifically based perspectives are translated into operational rules, policies and procedures through both big 'P' and small 'p' political processes. As the stage where the social judgments are made, this is often the most contentious. This debate encompasses a wide range of institutions and involves an array of actors, including scientists, policy advisors, public interest groups, industry and politicians. Ultimately, the political system is forced to make a Hobbesian choice. Although in a perfect world the risk system should approve safe products and technologies and reject unsafe ones, in practice we can never be certain. It is impossible to set up any scientific experiment to prove something is safe (because the possible harms are infinite). Hence, experiments are set up to test whether a specific technology or product will induce a specific set of hazards. Rejecting any hypothesis through experimentation would suggest the product or technology does not pose that specific risk. Ultimately, most regulatory systems identify a range of likely or less-likely hazards that could be related to a product and test for those. States inevitably have to choose between two possible errors from any system they create. They can either err on the side of caution and do as many tests and procedures as they could imagine, thereby raising the possibility that they will make a type II error, rejecting a safe and efficacious product or technology (more likely they would simply discourage the commercialization of a new product or process given the cost and delay inherent in such a system). Alternatively, they can accept a higher level of risk and undertake the most obvious reviews but quickly and efficiently make decisions about products. This raises the specter of them making a type I error, which involves approving a product or technology that is unsafe. It is impossible to cover both bases: a state must choose where to draw the line and what balance of type I and type II errors it can accept.[23]

There are a number of procedures that are used in most countries to try to balance the errors. In order to avoid type I errors, regulators usually adjust the allowable exposure thresholds by a factor of 10 (for example, if 10 parts per million would be safe, then only 1 part per million might be allowable) to cover the possibility of uncertainties in the hazard-exposure analysis. Similarly, there is increasing interest in adopting some form of precautionary approach (some call it the precautionary principle) that allows a regulator to delay or suspend judgment on a product that is suspected to pose unacceptable, irreversible risks but where the necessary science is not available.[24]

In practice, risk analysis is a heavily networked activity, involving a mix of public regulators, private firms and various values-based, interests-based and epistemic authorities. Although national governments have the power and authority to structure their own regulatory systems and national regulators have the legal authority to make decisions in the context of a legislated regulatory regime, most national systems are now heavily interconnected. The internationalization of both science and trade has led both to more coordination (gradual narrowing of differences between systems based on voluntary international codes of practice) and harmonization (standardization of regulation in identical form).[25]

The processes of coordination and harmonization have varied widely in pace and scope over the past century. The earliest and still the most extensive international collaboration has focused on human health related to food, drugs and chemical pollutants in the environment. One of the earliest efforts began in 1903, when the International Dairy Federation developed new international standards for milk and milk products. Over the intervening 50 years some commodity groups pushed for international safety rules to facilitate trade, but the general focus of most national governments was to establish rules and systems to protect their domestic industries and producers from external competition. After the Second World War, an array of new international organizations was created to coordinate and harmonize risk. The initial work was undertaken in specialist institutions created under the auspices of the United Nations. A few technical institutions like the International Office of Epizootics (OIE) predated this period but the creation in the 1950s of the World Health Organization (WHO) and the Food and Agriculture Organization (FAO) precipitated a significant effort to codify existing knowledge and develop international processes to further coordination and harmonization. These institutions, for instance, developed the Codex Alimentarius Commission in 1962 to coordinate efforts to standardize food labeling and safety standards. Over the years these organizations have grown to involve most of the countries in the world and hundreds of technical committees of regulators, scientists and industrial managers who meet regularly to consider new risks and to develop procedures for managing the processes of evaluating and managing those risks. These groups are for the most part closed, with admission controlled by the expert group in charge (be it national regulators, policy advisors or research scientists). At times the FAO or WHO have also commissioned expert panels of independent scientists to consider areas of uncertainty in the international food or health systems, seeking to contribute to greater understanding of how a risk or event might assert itself. Other regional and functional configurations of countries and experts have also engaged in the debate about regulating risks. The Organisation for Economic Co-operation and Development (OECD), created in 1969 by the largest industrial democracies, now has 30 member states engaged in an array of efforts to harmonize international regulatory requirements, standards and policies in the chemical and biotechnology areas. Apart from codifying practices and providing a forum for coordinating policy development, the OECD has created a new quasi-regulatory instrument in the biotechnology area called a 'Consensus Document'.[26] These mutually accepted scientific reports document the biology of a host crop plant or characterization of an introduced trait or transgenic method, thereby providing a common base to be used in a regulatory assessment of an agricultural or food product derived through modern biotechnology. Many of these are directly used in the assessment of new products in member states. More recently, there has been extensive bilateral effort among key producing and trading countries (particularly the Quad countries of the US, EU, Japan and Canada) to compare current processes and approaches in order to find a way to further harmonize the systems. These processes have generated a range of regional agreements, memoranda of understanding, mutual recognition agreements, formal dialogues and joint research projects.

Harmonization efforts got a major boost with the negotiation of the WTO in 1995. This agreement, for the first time, incorporates procedural standards related to the regulation of risk as part of the trade policy agenda. The overall agreement and its multitude of sub-agreements (more than 60 by some counts) are designed to contribute to a more liberal trading environment by reducing tariffs, preventing discrimination in trade, improving market access through limitations on the use of non-tariff barriers, requiring transparency and providing mechanisms to resolve disputes. In the context of risk assessment, the WTO permits national standards or regulations for the classification, grading or marketing of commodities in international trade (Art. XI.2.b) and for the adoption or enforcement of measures necessary to protect human, animal or plant life or health (Art. XXb). That authority is then delimited in the sub-agreements related to sanitary and phytosanitary measures (called the SPS Agreement) and technical barriers to trade (called the TBT Agreement). The SPS Agreement offers four rules on how that authority can be used: SPS measures cannot offend the principle of non-discrimination; SPS standards which conform to international standards developed by international organizations (such as Codex, OIE, and IPPC) are presumed to be consistent with the obligations outlined in the SPS Agreement; where a country implements national standards that are in excess of established international standards or where no international standards exist, those standards must be based on scientific principles and the completion of a risk assessment study; and SPS measures shall not be applied in a manner which would constitute a disguised restriction on international trade. The TBT Agreement, in contrast, deals with measures that may restrict trade but are justified in terms of some other legitimate objective, as laid out in Article XXI. Although measures under the TBT Agreement do not require risk assessments, they still must be non-discriminatory, must not be any more trade restrictive than necessary to realize a legitimate objective and must operate transparently.

The operation of these two sub-agreements has created a strong incentive for countries to conform to international standards, such as provided by the OIE, IPPC or Codex Alimentarius. In doing so, the potential for diverging scientific and procedural results is sharply reduced, as those scientific and technical organizations negotiate and establish harmonized international standards by consensus. This set of institutions operates by making it much more costly for a country to diverge from accepted standards. A range of decisions under the WTO disputes settlements mechanism – specifically related to EU measures on beef hormones, Australian measures affecting the importation of salmon and Japanese measures related to agricultural products[27] – have confirmed the range of measures allowable under the SPS agreement, sharply narrowing the choices for member states. In the EU hormones case, the WTO also granted the US and Canada the authority to implement retaliatory sanctions on the EU when it refused to change the measures to conform to the Agreement. Although authority still lies with the nation state to decide how to regulate, they can no longer export the costs of those measures abroad. Rather, they must internalize the costs of any deviating policy.

Meanwhile, there has been increasing effort to generate a similar level of interlocking international authorities and institutions in the area of risk and the environment. A range of negotiations have been simultaneously undertaken, leading to a variety of international agreements. Negotiations that began in 1982 at the Third World Congress on National Parks and Protected Areas in Bali, Indonesia, culminated at the 'Earth Summit' in Rio de Janeiro in 1992 with the Convention on Biodiversity (CBD), included as Agenda 21 of the United Nations Conference on Environment and Development. A number of international regulatory and policy efforts, some under the CBD and some independently initiated, now govern greenhouse gases and climate change (the Kyoto Protocol, 1997), traditional knowledge and plant genetic resources (the International Undertaking on Plant Genetic Resources, 1983, and the International Treaty of Plant Genetic Resources, 2001), endangered species (the Convention on International Trade in Endangered Species of Wild Fauna and Flora, 1975) and the transboundary movement of living modified organisms (Cartagena Protocol on Biosafety, 2000). Although each of these agreements involves processes that encourage and support greater coordination of policies and regulations related to the environment, none of them has the effective power of the WTO, with its dispute settlement system that has the ability to inflict economic pain on countries that do not conform. Each still relies on the best efforts of its members (and the US, for one, is not a formal party to any of the CBD follow-on agreements because it has been unwilling to ratify the original 1992 Convention).

Ultimately, risk analysis systems have to make decisions. When and how they make their decisions will partly depend on the nature of the proposed product and partly based on the decision-making structures inside the regulator. One point that many have noted is that 'regulatory science' differs in several key ways from academic science. Harvard science scholar Sheila Jasanoff has concluded that the regulatory system places particular stresses on what we know as science.[28] Knowledge claims presented in support of a new product, articulated in a positivist, deductive, hypothesis-driven set of empirically validated studies, are usually 'deconstructed' during the regulatory review process, to expose areas of weakness or uncertainty. At the end, regulators then have to 'reconstruct' the evidence to justify their decisions. In the process, they follow evidence and decision-making rules that are not supported solely by the underlying science. Although most of the national systems articulate goals, targets or deadlines for each stage of their review, few have adopted clear and transparent standard operating procedures that show the step-by-step process of assembling the evidence (what is required to be submitted and when), reviewing the evidence (who assesses the technical merits and how), making decisions (which decision framework is used) and resolving disputes (how are dissenting opinions accommodated).[29]

In absence of those clear guidelines, performance evaluations, audits and legal and trade disputes about the processes and their outcomes are far from conclusive. At one level, one can see some divergences between common law and civil code systems, where regulatory actions in common law systems tend to be more judiciable. In the US, for example, proponents (and with some limitations interested third parties) can challenge the processes and decisions of US regulators through the courts. In contrast, the European system was able to suspend review of a new technology for an indefinite period without any effective challenge (for example, even though the EU had a legally binding review system for GM crops, it did not decide on any new applications between 1998 and 2005).

Understanding how the current risk analysis framework truly operates and who is in charge is complicated because of the significant diffusion of responsibility and authority. In the first instance, the regulatory authority vested in national regulatory agencies is shared with other national agencies (often via mutual recognition agreements) and various international governmental organizations (such as the OECD, WTO, OIE, IPPC and Codex Alimentarius). By extension, many professional and epistemic communities, functioning under the auspices of one or more of these national or international institutions, are also heavily involved in defining the knowledge, procedures and protocols that make up the risk analysis system. In addition, proponents are heavily involved in any assessment and through them a variety of other actors are brought in. In the first instance, all regulatory regimes require owners of a product to be their own proponents. Although the regulators (alone or in collaboration with others) may define the variables of interest and the nature of the data needed to assess a product, the owner must undertake or contract with some third party to assemble the materials and data that the regulator will review and evaluate. In this sense, the owner is a key participant in the definition of the product and how it will be evaluated.

The regulatory review of Monsanto's recombinant bovine somatropin (rbST, sometimes called bovine growth hormone) intended for use in the dairy industry highlights the challenges facing regulators.[30] By the mid-1990s the company had developed using biotechnology as a process of mass producing a naturally occurring hormone that could be used to significantly increase the milk production of dairy cows. As one of the first major products of agricultural biotechnology to be assessed, it generated significant discussion among regulators and in the political sphere about how to deal with the new science of biotechnology. In the process, the regulators consulted regularly with the proponent and with others knowledgeable about both biotechnology and the dairy business, in order to identify the nature of the product, its intended uses and the evidence needed to assess its risk. Ultimately, the US government approved rbST for use in production and for consumption while Canada (and the EU and approximately 35 other countries) accepted the product as safe for human consumption but rejected it for use in their dairy herds.

The main difference in the two reviews seemed to revolve around the evidence gathered and assessed about when, where and how the technology was likely to be used. Monsanto argued, supported by a number of academic and industry studies, that rbST would be used primarily in intensively managed dairy operations, where cows are already producing large quantities of milk relative to unaided cows. In that context, the introduction of rbST in test herds did not generate statistically significant increases in animal stress, as measured by the presence of mastitis and other ailments. In contrast, the Canadian regulators required the company to undertake trials with both intensively managed and passively managed diary herds. Although the trials for the intensively managed herds delivered results similar to the US trials, the trials involving passively managed herds showed statistically significant increases in mastitis. After extensive debate in Canada, the Canadian Food Inspection Agency ruled that the new technology imposed unacceptable harm on the animals, thereby exceeding what was allowed in regulation. Hence, the technology was not allowed to be used on Canadian animals. In short, context is everything and, especially with transformative technologies, context can vary widely.

The end result is that most product assessments now involve a significant number of references to other authorities, with many nested decision-making systems contributing to the overall review. The challenge of such a system is that it is very difficult to show clear lines of authority, to document the standard operating procedures (often called SOPs) or to offer full and transparent processes. This system is being challenged on a number of fronts. The critique of the current risk analysis framework focuses on several considerations.

First, some are concerned that many of the most recent transformative technologies have not generated any significant political debate. Most of the big decisions on what, how and when to assess risks related to a new technology embodied in a new product are, or at least generally are designed to be, undertaken through a thoughtful and transparent political process, involving debate and decision by duly elected authorities. This has not been forthcoming. Rather regulators or the courts have simply muddled through, identifying and enacting incremental change in the regulatory system to handle each new risk as it has presented itself in the context of new products. One major concern is that there is a view, supported by some studies, that experts in the risk system are biased toward accepting higher levels of risk.[31] As a result, many citizens and civil authorities are suspicious that these technologies are in some way being furtively introduced, which offends their values of empowerment and involvement.

Second, a range of philosophers and ethicists have reasoned that there is little that is truly objective in the current system. Although a theoretical argument can be made, the most convincing argument often flows from case studies. Philosophers Conrad Brunk, Lawrence Haworth and Brenda Lee undertook a seminal study of the various regulatory decisions related to the cancellation of the registration of alachlor in Canada in 1985.[32] The chemical, an agricultural herbicide produced by Monsanto, was initially registered under the Pest Control Products Act in 1969 for use on corn and soybeans. In 1982 the regulator suspended the registration of the chemical and asked Monsanto to submit new long-term feeding trials for the chemical; the resulting studies showed that the chemical could induce cancerous tumors in laboratory mice. Over the next three years there was extensive debate and dispute about whether use of the chemical should be allowed. Ultimately, the product remained suspended. Brunk, Haworth and Lee examined the evidentiary base for the respective decisions and concluded that each of the key actors – the regulator, Monsanto as the proponent and a special review board – were forced to make normative choices when faced with incomplete information that could not be remedied by further scientific investigation. They identified five inherently normative issues related to the evaluation (what should one assume individuals applying the chemical would wear? would they use closed or open tractor cabs? should spills and other accidents be considered? should contamination of poorly constructed water wells be considered? when is an exposure scenario a 'reasonable' worst case?) and a series of conditionally normative issues (that is issues where there were uncertainties because there was not enough empirical data, such as which tests are most appropriate for determining exposure). The evaluation processes was also seen to involve value frameworks (is safety or economics more important?) which affected the participants' views of how to deal with uncertainty and where to assign the burden of proof.[33] Ultimately, the orientation of the various actors depended on whether they were trying to minimize type I errors (accepting a dangerous product) or type II errors (rejecting a safe product). The strong normative component in the process is a major concern to many people.

Third, there is concern about the nature of the decision frameworks used to evaluate risks and make decisions. Four basic approaches are used at various stages by various actors in the system: risk–cost–benefit analysis; revealed preference; expressed preferences; or natural standards.[34] Risk–cost–benefit analyses are framed in economic terms, using the Pareto criterion, which involves *ex ante* identification, monetization and balancing of risks, costs and benefits, where positive net balances would suggest proceeding and negative balances would indicate a proposal should be stopped. Generally the cost–benefit approach ignores the distribution of effects and any events that cannot be monetized (such as ethics). The two preference systems are attempts to identify and aggregate individual views about a technology. Revealed preference involves *ex post* examinations of the implicit valuation of life involved in different risks accepted by different individuals and expressed preferences are usually based on some form of survey or study of what people would consciously pay either to avoid a risk or to mitigate a risk. Finally, natural standards are frequently based on what is the background risk of an event (for example, a natural presence of radiation would imply a natural standard for acceptable exposure levels). There is no irrefutable basis for choosing one of the methods over the other. But choosing one to the exclusion of others will lead to both qualitatively and quantitatively different results.

Ultimately, each of these concerns contributes to a sense that the scientifically based risk analysis framework yields results that are increasingly diverging from the concerns and interests of individual citizens. Sociologists and ethicists have been at the vanguard of an effort to open up the review system. Many express the view that the only way for a regulatory structure to have any legitimacy in the eyes of citizens is for it to reflect the social valuation of the ordinal quality of the risks of a product and not just the cardinal estimation of the probabilistic risks. Risk analysis is fundamentally grounded on statistics and Bayesian decision theory, such that societal or public risk aversion is assumed to be a linear function of the average annual probability of fatality associated with the hazard. Lay people are seldom that ordered in their evaluations and views. In response, a wide range of scholars have proposed that the risk analysis framework needs to be fundamentally changed. One group calls for some form of citizen empowerment to overcome the perceived biases, limitations and normative elements in the system; this could involve advanced, informed consent (which is relatively easy with risks related to direct consumption but more problematic when dealing with risks that are unpriced externalities in the production or consumption of a product) or some form of more populist, open process (often involving citizen participation).[35] Another group calls for a change in the nature of the compilation and assessment of data, including use of alternative or competing assessments, more formal sensitivity analyses and different calibrations using a variety of ethical or methodological weighting schemes. Especially where the science is not definitive, as with most transformative technologies, there could be value in governments being presented with vigorous arguments in favor of various interpretations of both the scientific data and the risk assessment assumptions. One option would be to adopt the adversarial approach from legal procedure, with the proponent obligated to put forward both the strongest case for and the strongest case against using a new technology, which would then be assessed by the regulators.[36]

In a perfect world with no transaction costs and no opportunity costs any (or even all) of the suggestions may have some value but in the modern world with high transaction costs and clearly measurable opportunity costs each of the options would significantly rebalance the trade-off between type I and type II errors. Given that politics, the cornerstone of risk analysis, is the art of the possible, many of these ideas will need to demonstrate their feasibility before they become adapted and adopted into regulatory systems. The testing ground for many of those changes may be the evaluation of nanotechnology, which some assert offers both more opportunity and more threats than any previous technology.[37]

## EMERGING FROM THE GESTATION STAGE

The gestation stage is generally viewed as variable and uncertain. At any point a new technology, process or product could fail the tests of value generation, efficacy or safety. As noted, some technologies can move through these stages in a matter of days or weeks (such as computer programs) whereas others can take years (such as GM foods) or even decades (especially new categories of drugs). While in theory each of the three parallel systems uses different governing models and draws on different communities and actors, in practice the results from each of the processes has significant potential to influence other streams. New knowledge generated in the commercialization stream could influence what and how an invention is reduced to practice and the investigations and findings in the risk analysis system, even if they don't influence the regulatory finding, could influence what and how an invention is commercialized.

Even once a technology gets a triple green light – it has been organized into a potentially useful organizational structure (often as a patented technology or product financed by and directed towards a social or economic use), reduced to practice and scaled up, and accepted by regulators as not imposing unacceptable risks (most regulatory decisions involve letters of 'no objection' or some other non-judgmental assertion) – success is still far from certain. It is still necessary for an invention to be introduced into the market or society in a way that conforms to the needs and valuations of the end consumers and yet still generates adequate returns to justify its continued production.

## NOTES

1. On the ceiling of the Sistine Chapel in the Vatican.
2. Smith *et al.* (2003: 11).
3. For further discussion, see Phillips and Khachatourians (2001).
4. Mokyr (2002b) notes that although the earliest patents did not involve this feature, full disclosure became generally universal by 1770.
5. Chartrand (2006).
6. Committee for Study of Innovation, 2004.
7. This paragraph reports findings in a survey of researchers undertaken in the agri-food sector, reported in Phillips and Gustafson (2001). The applicability of these results in other knowledge-based sectors has been confirmed through subsequent informal interviews with industry and government laboratory managers.
8. Friederichs (2000) argues there is an array of potential strategies that often need to be initiated even before the invention is realized. The decision whether to patent, when and what to patent and where to file a patent on any inventions from a research program is often taken before the research effort begins. This is deemed prudent practice because patent laws require that the information related to an invention both be documented and be kept secret. Documentation is important because the US system adjudicates patent claims based on evidence of the 'first to invent'. Even though many other systems use the first to file rule, the dominant role of the US patent system and the importance of the US market forces most inventors to conform. The information in the documentation related to the research process and the actual inventive step, however, must be kept secret until a patent application is filed. Although inventors in the US have one year from first disclosure of an invention to file for a patent, almost all other countries require filing to occur before disclosure. Prior disclosure could either exclude the innovation from patent protection or enable another entity to appropriate the innovation and be the first to file for a patent on the innovation. Most parties to any research and development effort as a matter of form are now required to enter contractual non-disclosure agreements; these agreements have value in defining the nature of the relationship but, because of the cost of pursuing civil prosecution of any infraction, they are often ineffective if there is a premature disclosure and dispute.

9. Based on a survey undertaken in 2000 and reported in Phillips and Gustafson (2001).
10. Gold (2001).
11. Until recently, all information in a patent filing was kept secret until the patent was issued or abandoned. This at times led to a predatory practice of 'submarine' patents, where the inventor would file and then wait until someone else reduced the invention to practice and introduced it to the market. The inventor would then initiate the review, gain the grant and seek retroactive compensation from the unwitting infringing party. Countries over the years have shortened the period of secrecy, so that now most countries lay open the information in a patent application within 18 months of its filing, even if the patent has not been issued. A few submarine patents would appear to exist in the US system, but their number should increase as time advances.
12. This is consistent with Coase's (1960) assertion that if ownership was clear and bargaining costs low, then unconstrained private markets would lead to optimal use of property.
13. Dierker and Phillips (2002, 2003).
14. Hughes (1989).
15. Saxenian (1994).
16. Keiff (2003).
17. National Venture Capital Association (2004).
18. A number of recent studies have examined the risk analysis framework in depth, including Isaac (2002), and Phillips *et al.* (2006).
19. Codex Alimentarius (2000).
20. Codex Alimentarius (2000).
21. National Academy of Sciences (1983).
22. Isaac (2002).
23. Cranor (no date) argues one cannot 'minimize two variables at once'.
24. The first articulation of precaution was in 1982 when the United Nations General Assembly adopted the World Charter for Nature. Since that time the policy has been expressed in numerous international publications and agreements on environmental policy and has become a point of dispute between various countries. Principle 15 of the 1992 Rio Declaration on Environment and Development, states: '(i)n order to protect the environment, the precautionary approach shall be widely applied by States according to their capabilities. Where there are threats of serious or irreversible damage, lack of full scientific certainty shall not be used as a reason for postponing cost-effective measures to prevent environmental degradation.' In 2000, the Commission of the European Communities (EC, 2000) attempted to clarify the principle's application in its jurisdiction by issuing a communiqué describing guidelines for using it in a politically transparent manner. The communiqué stated: 'measures . . . must not be disproportionate to the desired level of protection and must not aim at zero risk . . . comparable situations should not be treated differently and . . . different situations should not be treated in the same way, unless there are objective grounds for doing so . . . measures . . . should be comparable in nature and scope with measures already taken in equivalent areas in which all the scientific data are available . . . the measures must be of a provisional nature pending the availability of more reliable scientific data . . . scientific research shall be continued with a view to obtaining more complete data' (EC, 2000: 19–21).
25. Davies (2002).
26. OECD (2005c).
27. WTO Appellate Decision (1998a, 1998b, 1999).
28. Jasanoff (1987).
29. CBAC (2002).
30. Mills (2002).
31. Shrader-Frechette (no date) reports that a group of assessors in the Netherlands tested the reliability of experts, using actual empirical frequencies of failures in seven subsystems in operating nuclear facilities in the US (using actual data gathered by Oak Ridge National Laboratories) to calibrate some of the probabilities assigned by a group of assessors in the Reactor Safety Study (also known as the Rasmussen Report or WASH 1400). They found that five of the seven actual risks examined were higher than estimated by the 90 per cent

upper confidence bound and two were lower than the 90 per cent lower confidence bound. None were within the confidence interval of the study. The Dutch authors of the study concluded that this confirmed that experts tend to be overconfident.

32. Brunk *et al.* (1991).
33. Cranor (no date) argues that 'the very idea of a risk assessment is normatively laden. A risk is a probability of an undesired outcome. Thus a risk is the probability of an outcome that will be unfortunate or undesirable from some evaluative point of view.' As such, most choices about the experimental design are normatively based.
34. Shrader-Frechette (1985).
35. For example Cooke (1990) and Shrader-Frechette (no date).
36. McKitrick (forthcoming).
37. For example, Mehta, M. (2006), 'Paradigm wars: a post-normal science approach to regulating biotechnology and nanotechnology' in Phillips *et al.* (2006).

# 12. Governing production, marketing and consumption

> There is nothing more difficult to arrange, more doubtful of success and more dangerous to carry through, than to initiate a new order of things. . . . Men are generally incredulous, never really trusting new things unless they have tested them by experience.
>
> **Niccolo Machiavelli**[1]

## INTRODUCTION

The general model of innovation introduced and examined in Chapter 3 posited that innovation involved creations which are introduced, absorbed into and persist in the economy and society. In essence, innovations generate value from their use and not from their creation or existence.[2] Gestating and introducing a new product, technology, process, concept or institution and ensuring that it perseveres and continues to add value is the ultimate hurdle for new inventions. Although many ideas can get part way through the system, only a small subset gets to be a candidate for use and only a small percentage of that is used in any appreciable way for any significant period of time. A recent study of 1091 Canadian inventions concluded that only 75 reached the market, of which 45 lost money and only six were judged to be blockbusters, with returns on investment above 1400 per cent.[3]

The problem, in a nutshell, is that even the most exciting and technically innovative invention may not match up with the needs, wants, desires and preferences of individual consumers, citizens and groups. So far most of the discussion and analysis of innovation in this volume has focused on the role of industrial, governmental and non-governmental actors as they consider what can be and should be developed and introduced. Although considerable effort is undertaken in the invention and gestation stages to attempt to sort out what individuals need, want or prefer – in their roles as citizens, consumers and social actors – the real work only begins once a new invention is offered into the marketplace. At that point, an individual, in all their complexity, is faced with a real choice, rather than some hypothetical selection. Even though it is possible beforehand to attempt to sort out likely from implausible options, in practice the results of those processes are often too narrow and removed from reality to truly reflect the prospects for a new technology, product or process. Although Machiavelli, quoted above, was generally referring to political change, the sentiment holds true for transformative technological change – experience matters.

The main difficulty is that real choices are almost always constrained optimums resulting from seriously boundedly rational situations. In practice choices are highly contextual – this makes it hard to predict how any individual might respond to a new option. We do have different methods for considering various facets of an individual, but we have yet to find a way to explain and predict whether an individual will respond as a consumer, as a citizen or as a member of communities reflecting specific values, interests or epistemic concerns. For the most part we are unable to predict with any degree of certainty which new offering will be taken up and which will be rejected.

How individuals have responded to the introduction of genetically modified foods represents a good example of the challenge. When scientists, regulators and corporate strategists first mused about the potential to use the array of new molecular tools that have come to be known as biotechnology (including gene amplification, gene splicing and the various 'omics' discussed in Chapter 2), they undertook a 'rational' process of thinking through the implications of the technology, including discussions about their efficacy, safety, commercial promise and socio-economic and ethical impacts. The processes led to heated debate among proponents and opponents of the technology but the results were ambiguous about which technological options individuals would buy, use, tolerate or oppose.

In absence of any products to present in front of consumers, governments, industry and various academic projects conducted surveys of public opinion. A Eurobarometer survey in 1979, for example, found that 35 per cent of the European public saw genetic research as an unacceptable risk and 49 per cent thought the same about 'synthetic food'. In 1985 the National Science Foundation discovered that a strong plurality (49 per cent) of Americans thought biotechnology would generate more benefits than harms and 39 per cent anticipated a less promising future.[4] Canadians, Australians and Japanese, among others, generally had opinions somewhere between the pessimistic European view and the more positive American outlook. Although surveys have been taken frequently over the intervening period, the views haven't changed dramatically.[5]

Although some assert that techno-pessimism is on the rise, the evidence suggests that a clear majority of people view technological change as beneficial, but they are discriminating about which technologies they think will generate value. In 2002 a series of surveys in a number of countries asked individuals whether they thought four key transformative technologies would 'improve our way of life in the next 20 years'. Public opinion in the US, Canada and the EU was equally optimistic about the Internet and, although mobile phones were generally viewed favorably, Europeans were significantly more optimistic than North Americans. In contrast, Europeans were less convinced of the long-term benefits of nuclear power or biotechnology.[6] Within these classes of technologies, there is a wide range of views about specific applications. With respect to agricultural biotechnology, for instance, less than 20 per cent of Americans are skeptical about its use in fields and our foods but between 30 and 40 per cent of Canadians and Europeans would be classed as skeptical.[7] When asked about specific applications of biotechnology, consumers generally will express enthusiasm for GM drugs (86 per cent of 25 000 respondents in 30 countries surveyed in 2000 were strongly in favor), modest support for genes moved from microbes or other plants to express for pesticides in plants (66 per cent in favor) and increasingly less support for transgenic modifications involving genes from other kingdoms or humans or modifications that simply increase the volume, reduce prices or increase the profits of farmers or multinational corporations.[8]

Differences in views have been attributed to a number of factors. Context is generally regarded as important. Taking biotechnology, for instance, there have been significant transatlantic differences on research policy and the adoption of particular applications. Federal funding for stem cell research is banned in the US but in some European countries, such as the UK, it is publicly supported. The US and Canada regulate GM foods based on the product attributes and only require proactive labeling where there is an identifiable nutritional or compositional change or allergenic property whereas the EU regulates the technology and requires mandatory labeling of all foods produced using the technology regardless of whether there are any identifiable attributes in the finished food. The US and Canada have led in the introduction and adoption of genetically modified commodity crops (with 56 million hectares under cultivation in 2005) but the EU has approved only a subset of GM crops which, though planted in five member states in 2005, were cultivated on less than 300 000 hectares in total. Meanwhile the public debate was significantly more prominently placed in the UK and parts of the European media than in Canadian or American media sources, helping to increase awareness of the issue and, some argue, enhancing the divergences of opinions.[9] Furthermore, Europeans by many measures appear to have a different perspective on the immediacy and value of their local environment. One obvious difference is the available land and the relationship of agricultural production to the environment. Europe has 40 per cent more people occupying one sixth of the land mass of North America. Meanwhile, about 25 per cent of Europe's land is cultivated but only about 11 per cent of North America's is engaged in agriculture and food production.[10] Agriculture simply has a relatively larger footprint and makes a greater impression on the European population than in North America. When individuals think of recreation in Europe, frequently it will involve accessing land that is also engaged in agriculture (such as rambling, camping, hiking and bird or nature watching), whereas in North America few people would consider using farm land for recreation (except perhaps for migratory bird or big-game hunting). Ultimately, the differences in opinion are a function of experience with their food chain, different implicit valuations of the environment and how the opportunities or concerns are addressed by regulators, policy makers and the media.

Once a new technology is in the market it is possible to test prior assessments of beliefs and preferences. Although marketing theory suggests that people move through a hierarchical set of stages in translating opinions into action (one characterization posits that awareness, interest and desire successively contribute to action),[11] there is increasing doubt about the relationships. Attitudes appear to be only weakly related to behavior, particularly in what are called 'repertoire markets' involving frequently purchased consumer goods like foods. Most theory tends to undervalue the role of existing allegiances to brands or producer groups and other buying habits and fails to clearly and concretely identify the economic calculus that consumers will undertake to determine the benefits and costs of new products. Empirical research on the attitude-behavior link generally produces low correlation between attitude and future behavior.[12] The introduction of GM foods has raised some interesting questions about how our opinions are formed and how they influence our active consumption decisions. One recent review of a range of surveys and experiments concluded that the only effective way to determine how consumers will act when offered a choice is to test it.[13] When consumers are presented with theoretical choices (whether constrained by a theoretical budget or not), they tend to indicate that they would need major price discounts to purchase negatively perceived technologies (such as some forms of GM foods) but in selected cases might pay hefty premiums for positively perceived technologies (such as some Internet applications). In practice, however, people are truly constrained by limited budgets, have limited time and capacity to investigate claims about individual products and generally indicate that they are often quite content to make boundedly rational decisions involving less than full and complete information (there is some evidence that personal utility is often optimized when consumers have a handful of choices, and that utility declines as choice both narrows or expands). As a result, when presented with real choices in a real setting, consumers often respond differently to how they might respond in surveys. In some instances, consumers who express concerns about a technology ignore any evidence on labels or in other media and simply purchase based on price or brand recognition. At least part of the reason appears to be a paucity of time, as most repertoire consumer purchases are made in a matter of seconds, rather than minutes. Complex signals are often ignored in that context.[14]

Ultimately, the challenge of governing production, marketing and consumption of transformative technologies involves putting an identifiable feature in a good or service that can be identified, evaluated by the ultimate consumers (or representatives who reflect individuals' needs and wants) and provided in some socially valuable, sustainable way (often defined as where the benefits match or exceed the overall costs). This effort, focused on the individual, involves a complex web of government laws and agencies, private for-profit actors and a widening array of civil authorities. Economist Paul Seabright has noted that every modern market economy involves:

> trade associations, money-back guarantees, the Hippocratic oath, the *appellation contrôlée*, inspectors of weights and measures, the convention of 'my word is my bond,' trademarks and brand names, the salesman's uniform, the training certificate hanging on the office wall, the opulence of a professional waiting room that soothes the visitor's anxiety about how much it is all going to cost with the subliminal assurances that nobody who makes this much money can be peddling unsound advice – all these are signals, more or less subtle, that what the seller has to offer is the best anyone can reasonably expect.[15]

The structuring of those systems warrants further investigation.

## THE NATURE OF USEFUL THINGS

Most of the literature about new technology frames it in the context of a consumer good. Although many new technologies do end up being transacted through the market and consumed by individuals, a wide range of other technologies is consumed collectively through public or community provision. Nevertheless, the economic framing of various goods and services helps us understand how different systems may have more or less capacity to govern aspects of a technology once it passes the gestation stage. Economists have two specific ways in which they frame markets. On the one hand they look at the nature of goods or services, examining how they are consumed. On the other hand, they examine the stages of concluding a transaction related to a good or service.

As discussed in Chapter 5, all goods or services exhibit a mix of characteristics which determine whether unaided markets can accommodate their production and marketing or whether some other assistance would be needed. Three key factors – rivalry, excludability and voice – combine to determine the nature of the products – public, private or collective – and the type of markets needed to supply optimal amounts of the product. As discussed, rivalry is a measure of congestion, which is a function of how many people can use a good at the same time before the utility of the consumption begins to decline. Rival products can only be used by a single person (such as food) – less rival products (often ideas, processes and public health) can be consumed simultaneously by a large number of users. The more disembodied the property, the less one is likely to face congestion. There is an almost continuous range of goods, from perfectly rivalrous to completely non-rivalrous. A second feature is whether it is possible to keep people from accessing and using the product, known as excludability. The degree of excludability is usually defined in terms of cost, which depends on the physical structure of the property, the cost of duplicating and using the property and the legal and social impediments to use. Highly excludable products usually occupy space. Consumption of disembodied ideas or services, in contrast, is often impossible to limit to subsets of a population.

Although many economists generally are content to classify products in the context of those two factors, recently there has been interest in considering how 'voice' contributes to value. If production or consumption depends on information provided by others, it is judged to have high voice content. In a knowledge-based economic system, this is becoming increasingly important, as both production systems and consumption depend on interdependencies of others using and contributing value. These three factors can be combined to categorize three relatively discrete types of pure goods (public, private and collective, club or common pool goods) and a multiplicity of hybrid goods with a mix of these three properties.

There are optimal ways of providing each of these pure goods. States generally are the only competent, capable entity that can produce public goods, markets are optimally structured to produce private goods and various civil authorities are instrumental in delivering common or pooled goods. At times one or more of these goods may be produced by one of the other actors, but over time there will be competitive institutional pressure for the governance of the production system to shift to the appropriate sector. In the context of hybrid goods, which exhibit mixed properties, there is often no obvious or even preferred institutional structure. Sometimes one of the pure types of actors – states, markets or civil authorities – can accommodate and manage the delivery of the good, but quite often some hybrid institution is needed, where the knowledge and expertise from two or more of the domains is needed to effectively and efficiently deliver the good. Although there is unlikely to be a great deal of path dependency in the institutional and governing systems for pure goods – as over the long run the most likely institutional system will come to run the system, regardless of who begins the process – there is likely to be a great deal of path dependency in the provision of hybrid goods. Whichever institution – state, market or authority – initiates the provision of a hybrid good is likely to have a significant, long-term impact on how it is governed and controlled.

A second way economists have characterized the governance of a good or service is through the process of concluding its related transaction. Although many public and collective goods are not subject to formal transactions, some of the features of transactions theory offer insight into even those arrangements. Economic theory posits that there are three types of goods: search goods, where consumers can visually identify attributes before consumption; experience goods, which require consumption to determine the attributes; and credence goods, where the unaided consumer cannot know the full attributes of consuming a good, at least for some period after consumption.[16] In practice, a single product could embody attributes that fit all three types of goods. For example, if one is looking for a tomato, one could 'search' through the bins and find one that looks good, smells ripe and is apparently free of insects or disease. Once a consumer takes it home and eats it, they experience the quality of the fruit, judging it based on a variety of subjective factors, such as flavor and texture. Ultimately, the utility derived from that tomato includes any longer-term benefits or costs of consuming the product, which become known some time after consumption. These could include some benefits (such as the nutrition of the fruit) or some costs (such as food-borne pathogens like *E-coli* or salmonella) which would become known within a few days, or less discernable health benefits (such as anti-oxidants) or toxic elements (often carcinogenic elements) that may have only a long-term cumulative effect on a person's health.

Markets for search goods are for the most part able to function efficiently based on simple market transactions – barter economies and street markets all thrive with little or no government intervention. Markets for products which exhibit more experiential or credence factors are often unable to function without some external governor. Product markets for experience goods where there are repeated transactions, such as for haircuts or for non-durable products, often can operate with only limited regulation, either ruled by the public or private sector. Markets for experience goods with infrequent transactions (as for consumer durables such as automobiles or houses) often require some additional structures to effect the exchange (such as brands or warranties). Credence goods (such as foods claiming health benefits) pose a much greater problem for markets. Although transactions may be possible in unregulated markets for credence goods, the absence of consumer knowledge severely limits the potential for an economically efficient outcome. These markets require some public, private or collective regulatory structure to address the need for credible information.

A wide array of market and extra-market rules and institutions work to ensure markets (and increasingly non-market authorities) are able to address the difficulties individuals face when they try to find the right mix of technologies, goods, services and institutions to satisfy their needs and wants. Local, regional, national and international public agencies establish the legal and regulatory underpinnings for production, marketing and consumption of most products. In some cases the rules simply provide general support for producers and consumers to make their own judgments. In other cases they explicitly regulate key production or marketing functions, identify acceptable from unacceptable practices and use the power of the state to enforce compliance. Regardless of whether the state intervenes directly, private firms usually will closely govern the production and marketing systems for their products, to the point that some manufacturers will specify how their product may be used and engage directly in auditing and enforcing their rules with consumers. Increasingly, we are seeing collective bodies emerge from values-based, interest-based and epistemic communities and setting the ground rules for governing new technologies, products, processes and institutions.

Ultimately, the system of governing production, marketing and consumption of new transformative technologies involves a range of sub-routines embedded in the knowledge, research and gestation stages, supplemented with a few strong, integrative institutions and rules at the production and marketing stage. Three specific concerns tend to emerge with any new technology. First, even though the gestation process is designed to weed out inherently risky technologies, some risk may remain. Some unacceptably risky technologies and products may emerge in the market and some inherently safe products may still present significant risk if used improperly. Second, most consumers and citizens face uncertainty as they attempt to get the right amount and types of information about new, transformative technologies. Ensuring appropriate amounts of truthful and useful information about new technologies is far from simple. Finally, once a product is in the market and in use, there remains the possibility that adverse effects may emerge and threaten those who use the technology. Each of these potentials needs to be managed. In each case, a blend of rules, institutions and structures is put into place by public, private and collective actors to manage the use of the product to ensure it continues to add value.

Technologies that no longer add value equal to or greater than their costs quickly die out. Sustainable technological change, then, hinges on the ability of these institutions to manage both the generation and distribution of value in such a way that it continues to be viable. Although a number of processes governing parts of the system remain closely held by governments and firms, increasingly this effort is being undertaken in a collaborative way, with public, private and collective organizations contributing to the creation, maintenance and enforcement of accountable, responsible and transparent (ART) standards.

## THE EMERGENCE OF A STANDARDS-BASED WORLD

Although governments still have a monopoly on enacting laws and regulations and firms are uniquely positioned to manage arm's-length impersonal transactions based on marginal pricing and full and complete contracts, transformative technologies often present an array of features that make it difficult for either system to operate in isolation.[17] Increasingly, we need some third party, usually some form of a civil authority, to engage in a process that further defines the nature of the product and its safe and effective use. Civil authorities often establish standards for quality of the product (such as food or manufacturing quality standards) or for interconnectivity of various components (such as for computer programs and computers) and frequently undertake enforcement of those standards to assure consumers of the efficacy, safety and quality of a new product. Although in theory the state or market could do this, often the interdependent nature of new technologies and lack of external validation of new things raises serious reservations about them. In response, there has been an increased role for standard setting efforts.[18] Ultimately, standards are collective goods created by collective actors.

Standards at root involve a set of informally or formally collectively acknowledged attributes that determine the quality and value of a technology, good or service for consumers. The Deutsches Institut für Normung eV (DIN), a contributor to the International Standards Organization (ISO), defines a standard as a 'document, which has been elaborated consensually and accepted by an acknowledged institution and which lays down for general and recurrent application rules, guidelines or characteristics for activities or the results thereof, whereby an optimal degree of regulation in a given connection is striven for'.[19] There are four main types of standards: product-based standards, which dictate the form of the end product; process-based standards, which determine how something is made or cultivated; behavioral standards, which specify how producers adhere to specific process or product requirements; and place-based standards, such as geographical identification used for wine and foods.[20] The one common feature is that all standards involve some sort of judgment about what content or processes will be acceptable in a product.

Standards are linked to regulations and private brands (Table 12.1). At the extremes, regulations and brands are separate governing systems and do not involve any collective aspects. In many circumstances, however, regulations, brands and warranties can become generalized and emerge as the common basis for governing production or marketing in an industry, region or product category. Standards can replace public regulation in three ways. Some standards involve publicly adjudicated statutory devolution, such as contracted-out audits or private enforcement of criminal laws (such as societies for the prevention of cruelty to animals often enforce local animal cruelty laws). Others entail voluntarily entered contracts (including self-regulation or service delivery). In some instances private, self-appointed authorities may supplant public regulation (such as civil litigation, credit ratings and NGO audits of government). At the other end of the spectrum, some branded products, such as aspirin, have become de facto standards for whole classes of products. Although both routes are used at times, most standards evolve from collective action among producers and consumers, with active support by government.

Standards emerge from one of four main processes. Some standards are unsponsored and simply emerge from common practice. The 'qwerty' layout on most typewriters and keyboards, for example, emerged in response to a technological need, good marketing and lock-in due to professional typing courses offered across the US – there was no single, identifiable proponent or author of the standard.[21] Other standards are sponsored by individuals or groups who possess proprietary rights to a set of specifications. For example, IBM at times established interconnectivity standards for its mainframe and microcomputers and Microsoft generally establishes interconnectivity and programming protocols for programs and datasets integrating with the Windows™ operating system and Office™ suite of programs. A third set of standards evolve from voluntary standards development organizations (also known as the committee solution) – these groups often exercise authority delegated from the state to enforce their rules. Finally, some standards are laid out in government regulation or legislation as mandatory conditions for production or marketing of a product or service.[22] Those collective processes that underpin the sponsored and voluntary standards are usually developed under the DIN 820 rules for the standardization process, which involves years if not

*Table 12.1 The interface between regulation, standards and private brands*

| **Public action** | | **Collective action** | | **Private action** |
|---|---|---|---|---|
| Regulations for public good purposes | Regulations based on standards | Commercial and private standards | Private brands act as standards | Private brands and warranties |
| **Driver** | | **Driver** | | **Driver** |
| Public good market failures | Common pool goods requiring voice; collective rather than firm or regulatory-based | | | Private firm-based profit maximization |

*Source*: Adapted from Smyth *et al.* (2004).

decades of consultation and negotiation among producers, consumers and regulators.[23]

Ultimately, standards-based governing systems are nested governing systems. As discussed earlier, the public regulatory systems involve both significant subsidiarity with different levels of government and various civil authorities and critical pooling of authority with international organizations. Within countries, most national governments reserve the rights to regulate for health and safety but, in practice, they devolve significant authority to regional and local governments to add additional health and safety regulations to address local circumstances and frequently leave most of the oversight and enforcement to local health and safety officers. Meanwhile, most countries have now acceded to a range of international agreements that normalize and harmonize standards, such as the Sanitary and Phytosanitary Agreement of the WTO, the Codex Alimentarius Commission of the Food and Agricultural Organization and hundreds of international standards organizations. Similarly, private companies have a broad reach, either through their in-house production or via international supply chains that they co-ordinate, through which they impose significant harmonization on standards related to both product quality and production and processing methods. As these accumulate, there has been a trend for international efforts to harmonize them in order to open markets. Most international standards, therefore, tend to be the compilation of various regional, national and sectoral standards. Although many have expressed concerns that industries and firms will fight to harmonize standards at the lowest level available, most international standards for new products or technologies (whether created through negotiation or imposed by national governments, firms or sectors) have tended to establish norms that are higher than what might already exist (that is above the minimum) and often near the top of the scale (hence raising concerns that standards have the potential to constrain rather than facilitate international trade). In spite of fears that firms will migrate to regions or countries with low standards for labor and the environment, multinational firms involved in commercializing new products or technologies have often been observed to insist on advanced economy standards whenever they locate in developing country markets (for example, equality of opportunity for different classes and genders of workers, environmental containment in the production system, significant skills or education in the work force and frequently higher wages than in local markets).

Standards can either impair or promote technological change, depending on how they codify practices.[24] Depending on how they are constructed, they can variously promote or limit compatibility and interface, set minimum or maximum quality or safety levels, enhance or restrict variety or expand or restrict the quality, volume and variety of information. Standards are not necessarily good or bad, but the implementation of the standard can have major impacts.

Immutable standards can lock in traditional technology, reduce variety and choice, increase market concentration, bolster monopolies or be open to regulatory capture. All would impede transition to new technologies. In contrast, if standards evolve with a technology or product category, they can generate faster uptake of new innovations which could generate network externalities and economies of scale, thereby facilitating trade and market growth. Effective standards can at times also reduce the chance of premature or inefficient lock-in, increase variety and lower transaction costs. Similarly, standards that set minimum quality or safety can be used to correct for adverse selection and for negative externalities. In essence, standards can either work to stifle innovation or can work to encourage follow-on invention and innovation.

## POST-MARKET MANAGEMENT

Once a new technology, process, good or service has successfully completed the gestation phase, it becomes a candidate for production, marketing and use. At that point, a diverse and highly networked set of government, industry and civil authorities begin to govern the post-release production and marketing risks, uncertainties and liabilities. In each case, the key actors, and sometimes a subsidiary hybrid created by two or more of the actors, have a significant influence on how the product is governed (Table 12.2).

Most markets that new technologies, processes, goods and services enter are already extensively governed. The state has an array of laws related to how products in specific markets can be produced and used (including environmental laws, planning and zoning rules and production quotas), a plethora of laws and regulations governing the structure and operation of firms and industries (such as rules relating to incorporation, reporting, accounting, taxation and labor standards) and a set of general baseline and specific rules to sustain or create competition in markets (related to advertising and labeling laws and anti-trust). Firms and industries similarly have a wide range of formal and informal systems that govern much of their individual and collective efforts to manage their financing, training, labor force, logistics, technology, production and marketing. Finally, there are more than 17 000 formal international standards in existence that govern much of what a firm can or can't do with their technologies, processes, goods and services. There is extensive literature about the scope and operation of each of these discrete systems.[25]

For the purposes of this study, the static structures are less important than how they adapt (or not) to the emergence of some new, transformative technology. In some cases there are processes for handling change. More often, however, transformative technologies do not fit within any of the pre-existing structures – new structures need to be developed. The efficiency and

*Table 12.2 Governing risk, uncertainty and liabilities by the public, private and collective sectors*

| | Risk | Uncertainty | Liability |
|---|---|---|---|
| Public mechanisms | Post-introduction restrictions on production and marketing, such as segregation | Consumer labeling laws to prevent fraud; mandatory labeling for key attributes; marketing orders and rules defining quality differentiation | Health, safety and environmental regulations related to surveillance; product liability and tort laws |
| Private mechanisms | Post-sale contracts defining rules for production or consumption | Voluntary labeling based on standards; brands and trademarks backed by privately managed identity preservation systems | Private warranties and brands backed by traceability systems |
| Collective mechanisms | HACCP, ISO, GMP and other product and process standards operated by industrial associations or consumer groups | Industrial or geographic standards, such as *appellation contrôlée* or geographic indicators | Industrial or supply chain-based traceability systems based on international standards and protocols |

effectiveness in developing such structures and the capacity of the resulting systems to handle any residual risks, uncertainties and liabilities will most often determine whether a new technology will have a clean launch and sustainable future and, if it does, which firms and regions will profit from the high returns offered by most new transformative technologies. Technologies that are unable to generate such systems often will fail or will be localized into a narrow range of regions, markets or uses.

Not every new technology triggers all the potential post-assessment concerns. Those applications that are commercialized in well established and structured supply chains that effectively service consumer demand often introduce few if any risks to the environment or human health. Others, due to their nature, do not trigger concerns about post-assessment liability or pose any need to address potential torts. Computer programs, for example, can be introduced with limited coordinated post-assessment management because they seldom have major risks or liabilities that transcend the transactional arrangements. Other technologies such as advanced instrumentation and some food and drug systems are introduced and used in highly organized systems by sophisticated users, limiting the need for outside governance. Generally, technologies exhibiting or introducing significant production or consumption externalities or interdependencies that cannot be accommodated or managed within any transactional arrangement or those that transcend any existing governance systems may require outside help. Most of these involve large groups of producers and consumers with widely varying levels of awareness, knowledge, capacity and interest in the technology and its impacts. In those cases, new governing subroutines are often urgently needed. The introduction of new technologies into the global agri-food system has been one of the few areas where a full array of concerns about risks, information asymmetries and uncertainties and potential liabilities have engaged policy makers in industry, government, academia and the civil societies. Virtually every consumer and citizen around the world has an interest in how these concerns are resolved.

## Risk Management and Risk Communication

Governments are perhaps most concerned about managing quantifiable, verifiable risks in production and marketing systems. Although the regulatory systems (discussed in Chapter 11) should have stopped any products with unacceptable risks from entering production or marketing channels, candidates for commercialization that emerge from those systems can still pose substantial risks to the health and safety of individuals, communities and the environment if the production and the use of the product are not carefully managed. Some products may be perfectly safe for their intended uses but unsafe if used improperly or in other circumstances. For example, electrical appliances designed to be used in 110 voltage systems found in North America would either be ruined or cause serious electrical outages in 240 voltage systems in use in much of the rest of the world. Similarly, some electrical appliances may be safe to use in dry conditions but will only be safe in wet areas if fitted with fuses or plugged into ground fault indicators (GFI).

Similarly, most production technologies have both safe, optimal operating environments and risky or unwise applications. There is some significant concern that clear-cutting old-growth forests may be environmentally unsound, but in some regions clear-cutting is arguably better silviculture practice than selective harvest of old-growth trees. In food production systems, most chemicals approved for use in the fruits, vegetables, crops, livestock, poultry and dairy production are safe if used according to the approved methods, but can damage the environment or introduce health risks in the resulting food if improperly applied (such as if applied before or after rains or during windy days or if applied immediately before harvest and marketing). Similarly, most over-the-counter and prescription drugs have safe applications and doses but if taken for extra-label purposes or without careful consideration of dosages and drug interactions could be rendered ineffective or dangerous. For all these reasons, governments remain involved in managing the risks inherent in products they have assessed, evaluated and approved for commercialization and use.

Governments are particularly involved in post-approval management of most of those sectors where the production system has a potential to harm the environment or where the resulting products exhibit either experience or credence elements. In the context of production, most regulators impose some ongoing responsibility on the owner of a technology or product to manage any known risks, preserving legal authority to investigate and adjudicate suspected or reported cases of failure to comply. For example, as part of an environmental impact assessment of any new production facility or resource development, government approval often entails a finite, measurable set of performance standards (such as release of pollutants in parts per million over a defined period of time), which may trigger criminal prosecution if exceeded. In those cases, they may perform periodic or random audits of the production system to ensure compliance or they may rely on exception reporting by neighbors, competitors, suppliers, NGOs or in-house informants. Substandard performance can lead to civil or criminal prosecution, at times resulting in fines, temporary or permanent shut-downs, imprisonment of negligent corporate officers and employees or forfeiture of the operating asset.

Governments around the world are beginning to recognize that in a world of globalized production and marketing, risk can spread more easily wherever there are gaps or disconnects in the regulatory regime. In response, there has been a concerted effort since the late 1890s, with a sharp increase in the last half-century, to link the various national systems of regulation for evaluating and managing health and safety through international accords. These efforts range from coordination (gradual narrowing of differences between regulatory systems through voluntary international codes of practice) to formal harmonization (standardizing regulation in identical forms).[26] A spectrum of types of international regulatory collaboration exists, including: encouragement of insurance markets, self-regulation and syndication of regulation; catchment area regulation; mutual recognition and toleration of jurisdiction shopping; extraterritorial application of national laws from key markets, such as the US and the EU; nested or delegated regulation; and formal harmonization of domestic laws with international accords. In those contexts, national regulators frequently engage in an array of activities. Most efforts begin with basic information sharing about the dimensions of current systems, insights into emerging issues and problems, foresight on new technologies and alerts about violations, violators and enforcement. Most efforts go beyond that. Often there is effort made to synchronize timing of regulatory decision making, enforcement and strategic priority setting, involving exchange of technical data and judgments (related to technologies and their characterizations, uses, risks and detection). Increasingly there are efforts to build regulatory capacity, including through joint training, policy development and advisory meetings.[27]

For those technologies, processes or products with extensive potential externalities and post-market risk, governments have recently adopted a variety of more prescriptive regulatory and policy approaches to attempt to limit their scale and scope.

Biotechnologies and related products are perhaps the most heavily regulated in the market. All prescription drugs are required to be controlled, with access managed through prescriptions written by appropriate physicians and dispensed by registered pharmacists. In some cases, such as for entirely novel or experimental therapies, approval is subject to the drug companies restricting access to targeted patients, sometimes through hospital or university controlled therapeutic or research programs or sometimes by requiring ongoing monitoring and assessment of the therapy while it is under way. Some of these requirements can require years of follow-up monitoring and evaluation before the drug is made more widely available. In the area of agricultural or food biotechnologies, the state has similarly imposed differential restrictions on the use of approved traits. Crops such as cotton and corn which express *Bacillus thuringiensis* (*Bt*), a naturally occurring soil bacteria with insecticidal properties, for the first ten years of use (and for the foreseeable future) were required by law to be managed by the owner to ensure the potency of *Bt* as an insecticide. Currently, that requires the owners (mostly Monsanto and Bayer) to implement refugia programs in areas planted with these seeds wherever 50 per cent of the cotton area is planted to *Bt* varieties. This requires either 20 per cent of the corn area is planted to conventional varieties that are conventionally treated or 4 per cent of the area is planted to conventional varieties that are not protected in any way. The logic is that if all crops were treated with *Bt*, target insects may develop tolerance to the pathogen but by leaving some fields untreated (or treated by other chemicals working through different pathways) a population of susceptible insects should be sustained. Although the state mandates these programs, the companies owning the technology are responsible for managing them. In the past they have tended to sub-contract that task to the seed trade, using pull-through contracts with farmers to specify the terms and conditions of usage of the seed. Farmers have been the front-line managers of the system. The owners of the traits have used a variety of audit models, contracting either with university researchers or consultancies to assess conformity with the rules. In most cases the results of the audits have been published either by the auditor or by the regulatory agency in an effort to offer some transparency to the system. Although many of those audits suggest compliance has been mixed, there is little evidence of whether the owners of the technology have used their contractual power to punish those producers or seed companies that have deviated from the contractual obligations.

As new traits with potential risk factors have been introduced into a wider array of food crops (such as industrial or pharmaceutical traits added to canola, corn or soybeans), a more direct approach of management has been developed. Increasingly governments in different countries are requiring owners of technology to develop and manage a special type of differentiation system called segregation. Bill Lin of the USDA defines segregation as the requirement 'that crops be kept separate to avoid co-mingling during planting, harvesting, loading and unloading, storage and transport'.[28] Segregation systems are used when potential food safety concerns exist over the co-mingling of the segregated product and all other like products. These systems are fundamentally different from the many new and evolving industry-led identity preservation production and marketing systems (which will be discussed below) – at root, IPPM is used to capture premiums and segregation is used to ensure safety. Unlike in many marketing structures, participation in a segregation system is not optional – any producer or firm involved with segregated products has to comply with rules that have been approved by the regulatory agency. The private firm will have the responsibility for developing the actual system and for managing it but the regulatory agency will be the final arbiter on approving the system for field use and the ultimate auditor of its performance. The focus of product delivery within a segregation supply chain is downstream. Although these segregated commodities commonly have industrial value and hence some demand-pull, it is less important that they meet those needs than that they remain contained. Testing and auditing are vital features of segregation systems and usually are conducted by agents of, or acting on behalf of, the regulator. Firms in major food exporting countries – including Australia, Canada, the EU states, New Zealand and the US – indicate that they are also working closely with their regulators to increase industry involvement in food safety, through industry-based standards systems such as Good Manufacturing

Practices (GMP) or Hazard Analysis Critical Control Points (HACCP) systems. The original HACCP program, based on the total quality management theories of W. Edwards Deming, was developed in the 1960s by the Pillsbury Company, the US Army and NASA as a collaborative effort to produce safe foods with 'zero defects' for the NASA space program. Pillsbury replaced traditional quality systems based on finished product sampling and testing with a new process that used operator control and continuous monitoring techniques at critical control points. Pillsbury extended that program into its commercial food production systems during the 1960s and publicized its form of the HACCP concept publicly at a conference for food protection in 1971. By 1974 the US FDA had adopted HACCP principles in new low-acid canned food regulations. Codex initiated in 1991 a working group to formalize a worldwide approach and application of HACCP principles. Most of the major exporting countries in the OECD indicate that they are pursuing HACCP-style risk-management in their agri-food systems.[29]

At first glance these systems would appear to be relatively conventional relationships between the state and key market actors but they are actually highly interconnected and distributed governing systems. The state has a clear and unambiguous role in setting the overarching goals for such systems and the proponent is ultimately responsible for implementing and paying for any system but, in the process of creating and managing the systems, a wide range of other actors are involved. Industry associations, private sector service contractors, academics and NGOs are often heavily involved in making the systems work. Industry associations often develop common protocols for use by their members (such as CropLife's stewardshipFirst™ agri-chemical program which provides a full set of procedures and protocols to cover the full lifecycle of most proprietary chemicals and its related Biostewardship module that manages GM seeds)[30] and various private sector service contractors are used by both regulators and proponents to enforce and audit the resulting systems (the USDA Grain Inspection, Packer and Stockyard Administration has authorized 53 different state and private partners to assist with standards enforcement, which currently involves around three million inspections annually). Meanwhile academics are occasionally called in to undertake audits (for example, faculty from the University of Guelph first audited the corn refugia program in Ontario and Quebec in 2001) and NGOs often engage in purpose-built monitoring of private sector performance.

There are very few open-field crop segregation systems presently operating in the global agri-food system. The best-known segregation system in Canada is for high erucic acid rapeseed (HEAR), which has industrial value due to the high acid content – there have been on average 11 HEAR varieties under cultivation in Canada in recent years. Two varieties of transgenic, novel oil canola

(Calgene Laurical™ varieties) were also contract registered (requiring formal segregation) and produced between 1996 and 1999 and there was a small amount of *Brassica juncea* segregated beginning in 2002. All pharmaceutical trials or production require full segregation. As one would expect with these types of systems, there is no tolerance for co-mingling. As a result, the costs have tended to be higher and, by implication, the premiums must be higher. HEAR earns on average a C$60/tonne premium in Canada and unsubstantiated reports peg premiums well over C$100/tonne for other specialty industrial and pharmaceutical crops. In the US, segregation is becoming more sophisticated as it adapts to the emergence of the first of what many suggest will be a wave of new pharmaceutical crops.[31]

When these systems work, they are virtually invisible to most citizens, consumers and social activists. The risks remain latent and not realized and most of us are content to concentrate on other concerns. Failure, however, generates significant debate. There are two possible failures in such systems. In some cases, undifferentiated products could enter the differentiated chain, which would likely reduce the value of the resulting segregated product but would not trigger any regulatory response. If, on the other hand, some of the differentiated product escapes the segregation system and enters the food and feed system, this would most likely trigger a complete recall of any products suspected of being affected and could result in civil and criminal prosecution in serious instances. Failure in this sense is not simply a lapse in anticipating all of the risks in an assessment (such as with thalidomide and Cox II inhibitors) but also includes those times when a technology may display acceptable risks when used according to design but unacceptable risks and harms when disbursed to unapproved users or applied in unapproved ways (such as cold medications containing pseudophendrine that can be easily reconstituted into illicit methamphetamines). In those instances, post-release controls have been increased to handle the risks.

In the agri-food world, the case of Aventis CropScience's StarLink™ corn illustrates the complex interactions between the state, market and civil authorities in governing post-release risks.[32] StarLink corn, expressing the Cry9C *Bt* gene, was developed by AgrEvo (subsequently merged into Aventis) in the mid-1990s and submitted for review to the US Environmental Protection Agency (responsible for assessing insecticidal technologies). It was registered and annually renewed for domestic animal feed and non-food, industrial use in the USA in 1998, 1999 and 2000. In order to satisfy EPA's objectives, Aventis established a 'Stewardship Program' that laid out requirements of contract registration for all fields of StarLink™ corn, the use of buffer zones and separate binning and targeted distribution of harvested corn.[33] Aventis distributed its seed through a variety of seed merchants and grain traders, requiring them to inform farmers and to enforce the segregation conditions.

According to Aventis, farmers planted StarLink on 248 000 acres in 26 states in 1999 and on 352 000 acres in 29 states in 2000.

On 18 September 2000 *The Washington Post* reported that tests commissioned by Genetically Engineered Food Alert, a coalition of consumer and environmental groups, found traces of genetic material from StarLink in Kraft taco shells in grocery stores in Washington, DC; that product was recalled within four days by Kraft. Subsequent testing revealed a much wider distribution of the genetic material – in November the FDA recalled more than 300 corn products. On 26 September 2000 Aventis instructed its seed distributors in the United States to stop sales of StarLink seed corn for planting in 2001 and shortly afterwards voluntarily cancelled its EPA registration of StarLink for feed and industrial use, thus taking the product off the market. A few days later, Aventis agreed with the USDA and EPA to purchase at a 25 cent per bushel premium all of the 2000-crop year StarLink corn, for an estimated cost of US$60 to $100 million. The fallout from the cross-contamination spread widely. In spite of agreements with the US regulators and key overseas markets to actively identify and withdraw the technology, small traces of the technology could still be detected in some part of the US corn business for a number of years. Aventis reached an out of court settlement with the attorneys general of 17 states in early 2001 for US$110 million to settle a class action lawsuit with affected corn producers. Ultimately Aventis sold its agricultural biotechnology business to Bayer (not including the liabilities for this failure). Some have estimated the costs of the failure for all concerned amounted to more than US$1 billion. Follow-up studies of this event suggested that the supposedly secure segregation system was fundamentally flawed, as the company never effectively audited the implementation by either the seed trade or farmers of its containment system and the regulator never undertook effective audits or enforcement. It was only through the efforts of the NGO community that the failure was reported and remedied. Beyond the financial losses, this failure had potentially significant long-term marketplace impacts. The inability of US corn exporters to guarantee their shipments as StarLink™-free enabled Brazil to export corn to Japan for the first time in 20 years.

Ultimately, risk management of new technologies is coming to involve a wide range of actors, including government regulators, technology owners, industry associations and the NGO community. Although creating and managing these systems may not require all of these actors, it is likely that the complexity and diversity of these arrangements may require the larger group of actors to fully audit and enforce the integrity of the systems.

## Truthful Marketing and Diffusion

In his classic textbook, Everett Rogers identifies five factors that he observes influence potential adoption of an invention: the relative advantage it confers;

its compatibility with current economic and social structures; its complexity; the ease with which it can be tested and tried by a potential adopter; and the ease with which an innovation can be observed and evaluated after a trial. Ultimately, these factors will vary depending on the ability to communicate. Rogers concludes that diffusion, thus, is 'the process by which an innovation is communicated through certain channels over time among the members of a social system . . . Communication is a process in which participants create and share information with one another in order to reach a mutual understanding.'[34] Diffusion of a new technology, thus, is fundamentally about communication within and among producers and between producers and consumers.

There is a large body of literature about how this process can be characterized. Rogers presents the results of a survey of 3890 articles on innovations and their diffusion drawn from 15 disciplines.[35] Typical innovations studied include technological ideas (such as the horse and boiling water), civic organization, agriculture, education, medicine and drugs, new communications and other technologies and new consumer products, focused on a range of units (including tribes, villages, individual farmers, firms, individuals and organizations) using a wide range of methods (including participant and non-participant observations, case studies, surveys, interviews, statistical analysis and model building). In aggregate, these studies identify the wide-spread relevance of the lazy S-shaped adopter distribution, the characteristics of adopters, a variety of channels of adoption, the spatial distribution of adoption, the roles of early adopters and opinion leaders and the consequences of adoption by individual actors or groups in society.[36] At root, the nature and pattern of the scale and scope of adoption can be traced back to how information is communicated.

The most critical actors in any adoption system are the early adopters. If the technology is being used as a new input into the production process, the firms that will first adapt and use new innovations are usually larger than average, more commercially oriented, operating in specialized market niches and having greater access to and making greater use of credit. When the technology is a final consumer good or service, the individuals most disposed to adapt and adopt new goods or services are better educated, have a higher than average social status, are willing and able to deal with abstractions better, are more disposed to change and are more socially linked. As one might gather from this characterization, there are no absolute measures to differentiate classes of adopters, but rather differences are more a matter of degree and relative importance of their bundles of attributes. In that context, Rogers found one way to characterize adopters using distributions based on the number of standard deviations a person's adoption is from the average time of adoption.[37] He divided the distribution into six classes. Innovators (the earliest adopters, sometimes alternatively called venturers, cosmopolites or gatekeepers) are those whose average adoption time is two standard deviations less than the average.[38] Early adopters adopted between two and one standard deviation faster than the average. Early majority or deliberate adopters adopted between one standard deviation and the average, the late majority or skeptics adopted between the average and one standard deviation later, laggards or traditionalists adopted later than one standard deviation after the average. In many instances, not all firms or consumers will be adopters, so the maximum adoption rate will lie between 0 per cent and 100 per cent, with theoreticians positing that the likely peak can often be calculated once the first inflection point in adoption is reached (that is, once the rate of increase in the total number of adopters begins to decline, which occurs when early majority or deliberate adopters begin to enter the market).

The shape of the adoption curve is fundamentally dependent on the communications related to the new technology, process, good or service. When credible, complete and timely information is available, the average lag in adoption may be shortened, the peak may be raised and, ultimately, the net present value of the invention will be optimized. The difficulty is that all new offerings inevitably begin with asymmetric information – the producer knows more than the consumer. If the products are search goods, then few problems are likely to arise. Markets should be more than able to get the message out, enabling different classes of consumers to search out the products and buy them. When the goods exhibit experience and credence factors, especially related to interconnectivity (that is how effectively an element will fit with pre-existing technologies, processes, goods and services) and long-term safety and efficacy, then unaided markets are unlikely to be able to achieve optimal uptake of their new invention. There is simply too much uncertainty for unaided markets to satisfy the information needs of consumers, especially average consumers who tend to be skeptical at the best of times.

In response, governments have made a variety of general rules related to marketing, product disclosure, advertising and labeling and at times engage directly in the management of a specific technology, good or service. On balance, the main goal of most government intervention in this area is to ensure consumers have access to an adequate amount of truthful information to be able to exercise their buying sovereignty while at the same time ensuring that firms engage in competitive ventures that optimize the economic benefits of markets. In that context, the general laws tend to be structured to ensure consumers have access to information about quality attributes, prices, appropriate method of use and any risk, adverse effects and possible remedies (both physical and legal) if the product fails to perform. Hence, most governments require labels on products, specifying the contents, weight, technical specifications, the name of the manufacturer and any experience or credence risks related to the product. The actual labels vary widely depending on the class of product. Electrical products do not need to specify the physical contents (for example, copper, steel or plastic), but do need to identify the appropriate use and technical features that would affect its utility (such as does it use 110 or 220/240 voltage? does it need to be grounded? is the plug compatible?). Food products, on the other hand, need to include a detailed list of contents, in decreasing order of importance, identify any unobservable potential allergenic properties, identify the full weight of the contents, and, in many jurisdictions, indicate the nutritional properties of the package or an average serving of the food. But food packaging often does not need to specify the individual producer (for example, fresh packed fruit can simply list the variety and not the individual grower) nor does it often require any explicit or detailed instructions for use. Drugs, on the other hand, require a two-level disclosure, with consumers getting abbreviated descriptions of the properties of over the counter and prescription drugs and with physicians and pharmacists being provided with detailed technical monographs on the properties of the drug. Each of these systems is built on the principle that information should at a minimum not be misleading or incorrect and, that where appropriate and possible, more information is better than less. Producers, meanwhile, are explicitly enjoined in law from promulgating misleading information about the nature of the product or the terms and conditions of its sale – for example, many jurisdictions prohibit advertising sale prices when the product has not been sold at its recommended retail price in the recent past or when adequate quantities are not available for sale at the discounted price. At times these provisions will be incorporated into advertising and marketing laws. At the extreme, unlawful marketing can lead to anti-competitive outcomes, which may abrogate provisions in anti-trust or anti-combines laws.

As with most forms of regulation, there has been a concerted effort in recent years to coordinate and harmonize these rules across markets and around the world. The World Trade Organization Agreement (1995) has become the focal point for most effort. The fundamental rules of commercial conduct were initially enshrined in the General Agreement on Tariffs and Trade (GATT) negotiated in 1947 and incorporated in its entirety in the WTO agreement in 1995. In brief, national regulations that might influence trade have to be in accordance with GATT principles concerning non-discrimination among countries in Article I (most-favored-nation treatment) and among domestic and foreign sources of supply in Article III (national treatment). In short, 'like products' should be treated equivalently. Article XI specifically permits regulations setting out national 'standards or regulations for the classification, grading or marketing of commodities in international trade' and Article XX(a) provides that countries can enact measures 'necessary to protect public morals'. Most importantly for labels, however, Article XX(b) allows the adoption or enforcement of 'measures necessary to protect human, animal or plant life or health'. To be consistent with Article XX(b), measures should not, however, be applied so as to cause arbitrary or unjustifiable discrimination between countries or act as disguised restrictions on trade. The Technical Barriers to Trade (TBT) Agreement of the WTO explicitly applies to technical standards used in trade, including labeling. Under the TBT Agreement, such standards must be non-discriminatory (Article 2.1) and must not be any more trade restrictive than necessary to realize a legitimate objective (Article 2.2). A wide range of legitimate objectives could justify labeling, including, for instance, prevention of deceptive practices, the regulation and maintenance of choice, the promotion of ethical treatment for animals or preserving public morals. Because there is no requirement within the Agreement for scientific risk assessment, the TBT Agreement could be much less restrictive than the SPS Agreement.[39] Nevertheless, the measure must be in response to an identified legitimate objective and must operate transparently. Governments are also encouraged to seek equivalence of technical matters and mutual recognition of conformity assessment procedures to reduce the restrictiveness of the measure.

Although at one level truthful communications about new technology can be viewed as two bi-lateral relationships – one between producers and consumers and another between producers and the state – in practice a number of collective actors are increasingly engaged in defining the rules for fair and full disclosure and competition. Many industrial and commodity groups, such as BIO, BIOTECanada and EuropaBIO in the biotechnology industry, have developed rules to govern communications and information dissemination about their common technology. Meanwhile a plethora of national and international standard setting bodies have been working to more narrowly define what different claims might mean and advertising councils and government regulators have been heavily involved in developing principles, protocols and rules to identify what is legitimate communication and what is unacceptable.

Ultimately, what is communicated must be fundamentally based on what is produced. In some product categories, such as computer software, virtually all of the core features are produced, defined and marketed under the aegis of firm or sectoral standards. In others, however, the state developed a range of market-making rules. In the agri-food world, for example, most countries have used articles XI and XX of the WTO to enact marketing orders that specify the range of acceptable inputs (for example, most seeds and farm chemicals are evaluated and registered by government agencies), define certain core production practices (for example, land use patterns are often restricted by local or national rules) and enact marketing systems that classify, codify, communicate and, sometimes, proscribe the marketing arrangements for certain qualities, volumes and prices of the resulting crops (for example, in Canada all wheat exported must be graded by the Canadian Grain Commission and marketed through the Canadian Wheat Board). Although private ventures may add to those basic rules and further differentiate their product, they usually cannot abrogate the general rules underpinning the system.[40]

Increasingly we are seeing traditional commodity-based markets differentiate as life-science companies, producers and manufactures seek to introduce and commercialize new innovations in the system. Although this is commonplace in most markets for manufactured products, arguably the biggest changes are occurring in the agri-food sector, where homogenized commodity production has been the main pattern of production and marketing. Now a combination of new production possibilities and new consumer preferences is leading to dramatic changes in the production and marketing system. Most of these new systems exhibit some form of identity preserved production and marketing (IPPM). These systems are almost always initiated by private firms in the food industry in an effort to extract premiums from a marketplace that has expressed a willingness to pay for an identifiable and marketable product trait or feature. An IPPM system is a 'closed loop' channel that facilitates the production and delivery of an assured quality by allowing differentiation of the commodity by preserving the identity of specific germplasm or breeding stock through the supply chain and ultimately into a processed product on a retail shelf.[41] The objective of an IPPM system is revenue management. Premiums need to be available to attract participants and the efforts of participants will be directed toward capturing a share of the premium. Participation in these systems is voluntary. The lead stakeholders in IPPM systems are private firms seeking to capture the increased value of special traits. The role of the regulatory body is to ensure that industry standards are handled in such a way as to prevent consumer fraud or anticompetitive industrial outcomes. Product information is often asymmetric, as only the product seller can know with certainty what level, if any, of cheating has occurred in the delivery of the product. Moral hazard may also be present due to the presence of premiums. Effective IPPM systems that span entire supply chains must have accurate two-way information flows. This means that information about purity and quality of the product flows downstream and information coming from consumers must flow upstream. Although the information flow in IPPM systems is two-way, the focus of the system is downstream. Each participant in the system wants to ensure they extract a portion of the value of the special trait, whether from production, processing or retailing the product. This means that each participant will focus on the needs of the next participant in the supply chain, much in the way that relay racers focus on receiving the handoff of a baton, running their leg of the race and then making a clean handoff to the next runner. Market failure, which can be due to lax communication or due to opportunistic behaviors, can (if particularly egregious) result in fraud charges for improper labeling and also create awareness with consumers that certain brand names or product categories cannot be trusted. Second parties, acting on behalf of the brand owner or developer of the special trait, will usually do testing and auditing.

Numerous IPPM systems are operating around the world. Some involve only the breeders and the wholesale market or processor and others extend through the entire supply chain to the retailer. Their structure depends on the attribute they are trying to preserve. Some novel oils, such as low-linolenic oils that are more stable in fryers, only have value at the processing level. Others, such as high-oleic oils, have health attributes that can be marketed to consumers. Identity preserved production and marketing systems are important for providing credence information to consumers about the provenance of a product, as those attributes are not visible or detectable in the product itself. A number of IPPM systems operate in North America. Although organic products are perhaps the most noticeable IPPM products, Cargill has an IPPM system in place to export canola to Japan (the variety gives off virtually no odor when used to fry food), General Mills operates an IPPM system for a select variety of white wheat that possesses a special trait for 'flake curling' when processed into breakfast cereal and Dow AgroSciences uses an IPPM system to export the Nexera canola variety to Japan where it is sold into the specialty gift oil market. In each of these systems, there are a range of costs and benefits. Essentially, the systems must generate adequate premiums in the intermediate or final markets to justify the added costs. The market actually determines the value of the systems, by revealing what tolerance for off-types it will allow (which sets the benchmark for costs) and then by revealing the incremental premium for the differentiated trait. A number of systems studied have been calculated to have incremental costs (over the commodity system) of US$20–35/tonne, and by assumption their continued operation implies that the premiums must exceed those levels.

These systems are almost always governed by a range of actors.[42] Governments provide the legal underpinnings of laws related to consumer fraud, labeling and contracts, while individual firms or collections of firms in a supply chain will strike civil contracts or relationships (a few may be able to rely on spot market transactions but most have some degree of formal contracting that specifies performance, rewards and penalties). Various collective organizations, including standards bodies, technical or sectoral industry associations, not-for-profit entities or profit-seeking ventures will undertake an array of functions in such systems, including standards identification (both consumers and producers usually need to be represented by collective actors in such systems), extension with producers (the potential for free riders often drives producers to pool their extension efforts), market making with consumers (free riding similarly drives this function into the hands of collective organizations) and auditing of performance (which can be done by any credible second-party auditor, such as an academic, registered certifier or industrial association or users' organization). The credibility and, by extension, value of the information flowing from such systems is only as good as the reputation and performance of the system in delivering what people want. In this limited circumstance, the consumer is always right. Whenever confidence is shaken, whether by an inadvertent or opportunistic failure, consumers will often take flight. Rebuilding confidence in a product line is expensive, time consuming and frequently ineffective.

## Liability Management

New technologies, especially new transformative technologies, are inherently unpredictable. Although they may generate significant benefits with minimal harms, there is always the possibility that some unanticipated harm will emerge with greater use. How the liabilities generated by those harms are managed can have a major influence on which technologies are developed, who can competitively develop new technologies, where those technologies will be developed and used and how fast and widespread the technologies will be applied.

As discussed in Chapter 6, one of the basic functions of the state is to draw the lines between actors in society by allocating rights and obligations and by enforcing those boundaries. Most governments operate two very explicit but somewhat different systems that clearly and, in many cases, effectively delimit the rights and obligations of technology developers and users. First, governments everywhere have some form of civil code or common law that explicitly or implicitly specifies that buyers and users of a good or service have the right to not be harmed by the use of a product. In short, the onus is on firms not to produce or market goods or services that can generate harms. In short, owners, producers and merchants bear some immutable obligations to stand behind their product. This provision has led to a range of different tort systems around the world. In some markets, such as the US, the right to tort coverage is enhanced by the ease of prosecuting a class action suit on behalf of any and all potentially harmed users, the ability of lawyers to work solely on percentage terms and the willingness of courts to award punitive as well as real damages. In these cases, firms that produce a technology that generates harms place at risk not only their sunk investments in development and commercialization, but can risk the entire company if courts determine the tort was egregious enough to justify large punitive damages.[43] Although most other countries have tort systems, many have much higher thresholds for the certification of a class action, many prohibit lawyers from working solely on a percentage basis and most have limits on punitive damages or on the scale of damages that can be assessed.

Second, although many new technologies are evaluated, assessed and 'approved' by government regulators, governments never assume responsibility for the quality of the product once it is in the marketplace. Drugs, for example, are provided with letters of 'no objection' and foods are deemed to be 'generally regarded as safe'. Government regulators very explicitly assert in their rulings that all of the liabilities related to a new product lie with the owner or proponent. Although regulators can be liable if they fail to undertake their reviews according to the procedures and protocols specified in their national laws and regulations, they usually are absolved from sharing any liability resulting from unanticipated or unknowable consequences. Firms that produce and market a product that generates harm ultimately are liable for any resulting claims.

The operation of the for-profit legal system and various values- and interest-based collective organizations make the tort system function. In practical terms, individual consumers or users seldom have the capacity to adequately identify and define the nature of a harm. Nor do they have the resources to launch and sustain a tort action against most firms. The challenge is simply too great for individuals to have much hope of success. In many countries the legal system and various civil authorities have stepped into the breach. Product liability litigators everywhere are ready and willing to entertain proposals from people who believe they have been harmed by some product in the marketplace. How aggressive they can be is often determined by the terms and conditions for their involvement. In some jurisdictions, these types of actions are viewed unfavorably within the legal profession and in society more generally; in others this type of action is viewed as legitimate and valuable. Similarly, in many jurisdictions, firms must have fee-paying clients to launch a case; in others lawyers may work for free, in anticipation of sharing a larger reward at the end. The professional codes of conduct in the Benchers and other legal organizations have a major influence of how these and other factors will be handled. Meanwhile, the role of values- and interest-based civil authorities ranges widely. Many of the most significant class action lawsuits in the US were coordinated by professional managers and campaigners in key values-based NGOs. They have the reach, capacity and authority to identify harms, work with firms and regulators to handle those amendable to regulatory action and to commission lawyers to prosecute cases. Some groups, such as the Sierra Club, have actually created their own legal firms to further their direct action efforts. A range of general consumers, product users and patient advocacy groups have been similarly engaged in using the tort system to improve the auditing and enforcement of various products.

Overarching the entire liability system are the private equity and capital markets. They make the actions in the tort system more binding on firms. In the first instance, state and national regulations require companies to report all items of material impact on their firm's performance, including any legal actions that could affect revenues, costs or equity. Only a foolish or bold company, municipality or state issues debt without first getting a rating from Moody's, Standard & Poor's or Fitch. Standard and Poor's, the world's largest credit rating firm, rates $30 trillion of debt, representing nearly 750 000 securities issued by more than 40 000 borrowers.[44] Moreover, the Basel 2 Accords on bank capital requirements, effective in 2007, require regulators to use such ratings in assessing bank capital adequacy. Many pension funds and other investors also stipulate that their bond investments must have a rating from a 'nationally recognized statistical rating organization', the US Security and Exchange Commission's designation. The SEC recognizes only five agencies: Moody's, S&P, Fitch, AM Best (a firm that specializes in the insurance sector) and DBRS, a Canadian firm. Hence, a company's capacity to borrow could be significantly influenced by any outstanding liabilities and their ability to manage their costs. When an institution decides to issue debt, it calls a rating agency (or multiple agencies) to review the data and management and rate the creditworthiness of the institution. If the issuer objects to the conclusions, some agencies will allow it to veto the release of the initial rating. Ratings are normally reviewed annually but can be accelerated if conditions warrant. The rating is paid for by the issuer, which creates a concern about conflict of interest.

In the face of potentially catastrophic commercial harm if a new technology generates harms, firms respond in a number of ways. In the first instance, many established firms become conservative, preferring to pursue iterative technologies rather than transformational ones. This at least partly explains why many transformative technologies are brought to the market first by new market entrants rather than by market leaders. Most leading firms, however, recognize that the short-term gain of such a strategy may lead to long-term losses. In response, firms are continually seeking strategies to limit their exposure to liability.

One strategy that has been used for years in the area of manufactured consumer durable goods is the development of proprietary brands and company-backed or industrial warranties and money-back guarantees. These strategies work well for experiential and repertoire goods, but are often incapable of dealing with the credence elements of novel products. In response, firms seeking to manage their liabilities have invested heavily in traceability systems. The International Organization for Standardization (ISO) has defined traceability as the 'ability to trace the history, application or location of an entity by means of recorded identifications'. This is relatively easy for durable goods, where unit and lot numbers are relatively easily traceable but is much more difficult for goods such as food which are comprised of a wide array of blended and pooled primary materials. In the context of food, the Codex Alimentarius Commission has adopted the ISO definition as their working definition for all Codex standards. Traceability systems are designed to ensure that products available for consumption are as safe as possible and that harmful elements can be identified, isolated and removed from the system. Participation in a traceability system can be voluntary, depending on where in the supply chain the participant is located. The closer the participant is to the start of the supply chain, the more likely it will be that participation is voluntary. The lead stakeholder may be a commodity group demanding greater clarity in or selection of food products, a standards council that is comprised of industry representatives from all sectors of the supply chain or the regulator seeking to ensure consumer protection. Traceability systems encourage real-time, two-way flows of information in order to ensure the system can react quickly to food safety concerns. If a product is discovered to exceed any defined tolerance level at any point in the supply chain, traceability will be used to identify the source of the problem and to locate any and all retail and intermediate products that may be affected. Information on food safety flows upstream while information on specific products flows downstream. This results in the focus of traceability systems being upstream. Testing and auditing in such systems will usually be conducted by third parties according to the standards developed by third-party producer or consumer organizations.

There are an increasing number of traceability systems operating in global markets. Perhaps the oldest and most developed system is the manufacturers' lot number system that traces processed foods from plants to consumers. Those were implemented largely to handle liability of contaminated foods. More recently, UK retail chains have adopted extensive traceability systems (from retailer to farmer), initially for beef but now for a growing list of products, in order to manage their liabilities flowing from the UK Food Safety Act, 1990. More recently the livestock industry in as many as 25 countries has either introduced or announced plans to introduce partial traceback systems for cows (from slaughter house to birth) and there has been significant discussion about if and how a traceback system can be developed for commodity grains and oilseeds. Although these systems, which add between 1 and 5 per cent to the cost of a product, do not add any directly identifiable premium to the product, they do act as a form of insurance against a food contamination that could cause the recall of an entire product category. The recent detection of BSE in a Canadian cattle herd was estimated to cost the industry $15 million per day in lost revenues. The incomplete traceback system operated by the Canadian Cattle Identification Agency at that period will have paid for itself if it can reduce the trade embargo by even a few days. Even though no traceback provisions exist for GM foods yet, the EU has passed but not yet implemented a new set of rules that require traceback systems both within the EU and in the exporting countries, in order to validate claims of GM or GM-free status in products without detectable rDNA.[45] Once it becomes operational, this will be a novel use of traceback, as no other systems operating are imposed by the state – all of the others are economic choices adopted by liable firms seeking to limit their exposure to losses.

## CONCLUSIONS

The commercial challenge of introducing and sustaining the profitable and socially beneficial use of a new technology, process, good or service is daunting. Ultimately, the more novel and transformative the underlying technological innovation, the more effort will be needed to assess and manage risk, to differentiate the products to meet consumer uncertainties and to manage any potential liabilities flowing from the inevitable miscues of premature introduction or incomplete knowledge related to use of a new technology. The greater the challenge, the less likely that any individual firm or government agency will have the capacity, competence or authority to be effective. Rather, as the challenges become more complex, so do the likely solutions. The scope and scale of the problems are likely to differ depending on the context of the new technology and the pre-existing market and non-market circumstances but a few features are likely to remain common. It is important to keep in mind that adoption and diffusion are not comparative static exercises. The markets for both complementary and competing goods will continue to change during the period of introduction and diffusion. Producers of many existing technologies and products will strategically respond to new market competition, either by accelerating productivity growth or by lowering the monopoly rents they extract from the market, in order to remain competitive.[46] This type of 'last gasp' response can extend both the life of older technologies and products and elongate the diffusion process for new technologies and products.

The more technologies and their related products embody experiential or credence factors and the more they present externalities or interdependencies, the more likely complex, distributed governing systems will be required. Many of the functions of successfully launching and managing a new technology or related product will necessarily remain in the purview of for-profit firms, but other actors are needed to manage post-market risk, to resolve information asymmetries and to control potential liabilities. Neither private firms nor public agencies have the necessary capacity, competence, authority or trust to do the job alone. As discussed in Chapter 9, however, engaging civil authorities more in the system presents its own set of challenges, as they often seek incompatible goals and objectives and have different standards and expectations of accountability, responsibility and authority.

# NOTES

1. Machiavelli (1513: 51).
2. Committee for Study of Invention (2004).
3. Astebro (2003).
4. http://www.americans-world.org/digest/global_issues/biotechnology/biotech1.cfm.
5. Gaskell *et al.* (2005).
6. Ibid.
7. Ibid.
8. Environics International (2000).
9. Marks and Kalaitzandonakes (2001).
10. Calculated by author from statistics derived from CIA Factbook and FAOSTAT.
11. Strong (1925).
12. Phillips and Corkindale (2004).
13. Smyth *et al.* (2005: Chapter 4).
14. Einseidel *et al.* (2000).
15. Seabright (2004: 85).
16. Tirole (1998).
17. Bristol-Myers Squibb Company's website posting on corporate social responsibility shows the array of different approaches and standards setting systems one company could participate in (www.bms.com/static/ehs/linkst/data/linkst.html).
18. As discussed in Chapter 8, Power (1994) has labeled this the rise of the 'audit society'.
19. DIN-EN 45020 (edition 1994, p. 12, in DIN, 1995, p. 276).
20. Bingen and Busch (2005).
21. David (1985).
22. Blind (2004) cites David Greenstein's 1990 categorization of standards.
23. Blind (2004).
24. Blind (2004).
25. The public administration literature (for example Doern and Reed, 2000; Vogel and Kagan, 2004) offers insights in the regulation of the modern marketplace, the business school literature is heavily engaged in understanding how firms operate and govern their own structures and a number of economists and lawyers have begun to examine the role of civil authorities in managing markets through standards (for example Blind, 2004; Josling *et al.*, 2002).
26. Davies (2002).
27. Perri 6 (2001: 95).
28. Lin (2002: 263).
29. Codex (2000).
30. Hepworth (2001).
31. Smyth *et al.* (2005).
32. See Segarra and Rawson (2001) for a full background on the case.
33. Harl *et al.* (2000).
34. Rogers (1983: 5).
35. Rogers (2003).
36. Rogers (1983).
37. Ibid.
38. If, for example, a population of consumers of a specific product made their first purchases of the new product in a relatively normal, bell shaped distribution from immediately upon release to one year later, the average could be 6 months, with a standard deviation of 2 months. Using the average minus two SD would imply that innovators adopted on average 6–2(2) or 2 months after introduction, early adopters between 2 and 4 months, deliberate adopters between 4 and 6 months, skeptics between 6–8 months and laggards or traditionalist after 8 months.
39. Sheldon and Josling (2002).
40. Kennett *et al.* (1998) offers an interesting example from Canada where a foreign buyer (Warburtons from England) worked with a local grain company (Manitoba Elevators) to

develop a more highly differentiated production and marketing channel for specific varieties of CWRS bread wheat to operate inside the wheat pooling system.

41. Buckwell *et al.* (1999); Lin (2002).
42. The organization Business for Social Responsibility has produced an illuminating list of external standards setting and auditing processes for environmental performance – including the Business Charter for Sustainable Development, the ISO 14000 series, the EU's Eco-Management and Audit Scheme, the CERES principles, the UN Global compact, the Global Reporting Initiative, the Global Sullivan Principles and the Natural Step Framework (source: www.bsr.org).
43. The ease of tort cases in the US is almost legendary, with apparently frivolous cases being brought on behalf of defendants who have been injured or killed while using products in ways never envisaged by the producer.
44. *The Economist* (2005). Special report: Credit rating agencies: who rates the raters? *The Economist*, 26 March 67–9.
45. EU (2001).
46. Hall (2004).

# PART V

# Implications

# 13. Conclusions

> It is not the strongest of the species that survive, nor the most intelligent, but the one most responsive to change.
>
> **Charles Darwin**

## INTRODUCTION

Any discussion about innovation is ultimately about change. Although the rate, scale and scope of the change will vary depending on whether the technology involves small, iterative adjustments or poses large, transformative modifications, the challenge remains the same. Institutions need to respond and to adapt to the new circumstances. Although many institutions involved in governing a system are able to adapt to changing circumstances, inflexible institutions can block a change (for better or worse) or become increasingly irrelevant as new institutional pathways are developed to govern new circumstances. Existing institutions will end up either leading, following, or getting out of the way (either by design or by happenstance). The nature and process of making that fundamental choice about how to engage goes a long way to determining what, where, when, why, how and by whom new technologies are adapted, adopted and used, and whether that use generates social benefits.

This book has attempted to illuminate the range of potential changes that could be precipitated by technological innovation in modern, industrial economies and to examine how different governing actors have engaged in controlling change in different sectors and regions. As with all social analysis, this effort has been directed to discovering a deeper understanding of how society – through the state, market and a range of civil authorities – makes choices about what it will nurture and promote and what it will discourage or block. As Darwin suggests, it is not the strength of the system but rather the flexibility and adaptability of the system that will determine its lasting influence. If a system is perennially unresponsive, or blocks fundamentally sound and advantageous change, it then will face competitive pressures, either from within (for instance, society or some key actors may demand action) or from without (often resulting in declining economic or military influence).

This analysis is designed to provide a theoretical, empirical and policy context for the pressures governing systems now face and to assess the likelihood of different institutions being able to lead, follow or get out of the way. Theory and evidence are most persuasive when they can be directly used to anticipate change, to forecast the nature of change and to evaluate the impacts of alternative responses to change. In this sense, the more theory is behaviorally based, the more it can provide illumination. This book has attempted to ground our understanding of how the state (through its underlying assumptions about humankind), the market (based on individual optimization of welfare through transactions) and civil authorities (centered on alternative world views) will variously respond to pressures for change.

This concluding chapter examines some of the implications of this approach and offers a few observations on areas for further investigation.

## GOVERNING TRANSFORMATIVE TECHNOLOGICAL INNOVATION

The debate about who is controlling the rate, scale and scope of change pervades much of public discourse today. Discussions about the economy, our health and the environment, for example, are largely being driven or shaped by new opportunities or threats resulting from emerging technologies. As a result, analysis and discussion about the nature and scope of technological change is reaching ever increasing levels of hype, concern and angst.

Different technologies in different subject areas generate a variety of responses. The relatively small but influential number of technophiles in the world, disproportionately but not totally located in the for-profit business sector, are ecstatic about the opportunities that abound when new technologies, products, processes and institutions emerge, destabilizing the existing market relationships and presenting potential windfall gains. The plurality (or maybe simply the silent majority) of us are techno-skeptics, hoping that some of the hyped opportunities are credible and can be realized but cautious about the painful short-term adjustments and potential long-term harms that can accompany change. Finally, a small but increasingly vocal and mobilized group of technophobes are pressing to alter the distribution of power in order to change the trajectory of change. This group either seeks to redirect, halt or reverse the direction and rate of change or to restructure society in a way that different actors will contribute to, participate in or benefit (or be harmed) from changes. The distribution of those groups in society is for the most part a function of the nature of the new technology or innovation under consideration, the underlying context for the introduction of that change, the prevailing distribution of winners and losers from similar or past changes and expectations of future impacts on the fortunes of those groups.

As discussed in Chapter 2, technological change can vary quite widely. At one extreme, most technological change involves small, incremental, infill inventions that fit well within existing systems governing production, marketing and use. They emerge, are adapted and are adopted almost without any apparent effort or conscious decision by existing governors. At the other extreme, transformative innovations are much more dynamic and expansive. This type of change involves disjointed, step adjustments in our productive and institutional capacity which displaces, destabilizes or overturns pre-existing systems. They are usually driven by new epistemologies, they offer significant complementarities (within and across sectors) and they tend to involve convergent, recombined or hybrid technologies. Although they are often enthusiastically announced, they seldom remain under the control of any single sector, market or domain, as they tend to infiltrate and influence others over a long and variable timeline. The pervasive and expansive nature of transformative change creates uncertainty – it is seldom clear until well after a launch where, when, how, by whom and why the technology will spread to other uses – which tends to generate significant debate and controversy.

The complex nature of transformative change leads to much more extensive innovation processes, which involve a much wider array of actors. As discussed in Chapter 3, translating an invention into a socially embedded innovation involves a complex web of principals, agents, promoters and regulators on the supply side and middlemen, marketers and consumers on the demand side. Constructing new markets for new products or services is seldom straightforward or simple. Transformative changes thus mobilize a much wider range of actors, many who have never before expressed concerns or interest in change. Modeling this environment is difficult. Although the full array of economic, political and social connections could be modeled in theory, in practice only a relatively small subset of any interactions can realistically be incorporated. Recent efforts to model the soybean supply chain in the US and the intellectual property and biotechnology regulatory regimes in Canada involve hundreds of stylized relationships, behavioral assumptions and estimated interconnections.[1]

Ultimately, the transcendent, expansive nature of transformative technological innovation poses some real challenges for governors. This is one of the more significant, identifiable underlying concerns about globalization. Increasing interconnectivity caused by transformative technologies is undercutting both our perceptions of and our capacity to isolate and specifically control new developments. Information technologies, combined with the many and varied advances in biological and materials processes, have linked previously disconnected parts of our world both physically and emotionally.

## WHAT'S AT STAKE?

At one level, one could conceive of the challenge of transformative change as simply a fight between elites. One can argue (as many do) that power systems are controlled variously by governments, firms or social groups, depending on the endowments, capacities and institutions created in the past. Those who have power usually believe their power is both just and efficient. Innovation can either strengthen the position of those who have power or can undercut their position by introducing a new set of actors into the power system. Although a few prospective changes may be inherently democratic in nature and impact, most simply reallocate power among a relatively narrow group of actors. State and market actors in particular seek to monopolize power for their own use. Civil authorities do not necessarily seek overt power but the successful ones almost always need to apportion and mobilize some power to have effect and sustain themselves. In short, change could be viewed in Marxian perspectives as a fight among a chosen few.

Although elite power relationships may be at the root of many of our systems, how they operate (that is, who has and uses power) has a major impact on the lot of the rest of us. Our futures are linked to how transformative innovations are handled. Individual and national welfare are a function of the long-term growth and development of our productive capacity. Both absolute and relative changes are important. In absolute terms, our ability to introduce new ways of doing things that either increase our socio-economic capacity or reduce the material costs of our consumption will be the basis for our collective prosperity. One might reasonably argue that people living in OECD countries have (or should have) enough resources to live comfortably but the majority of the world's population has very limited access to resources and suffers from the attendant risks and uncertainties of inadequate food, shelter, clothing, medical care and public security and environmental safety. New technologies and new institutions offer some hope that they can attain a higher level of prosperity and security without having to undergo all of the trials and failures experienced by OECD countries. Furthermore, OECD countries are recognizing that their long-term security and quality of life may be linked to the plight of the underdeveloped parts of the world. The global environment and public health have been shown to be great equalizers in recent years. Adverse weather and environmental change have punished both developed and developing countries in recent years and even though most of the new public health threats – such as AIDS, antibiotic resistant infections and various influenzas – have emerged from underdeveloped regions of the world they have had major economic and social impacts in many OECD nations.

Equally important for many is the fact that the relative ability of countries to benefit from new technological and productive opportunities determines a state or society's economic, social and political status in the world. Countries that effectively and efficiently generate, gestate and commercialize new technologies and innovations tend to accumulate more wealth and power in the long run. Across time, one can see that countries that fail to adapt to change tend to ossify and lose status and power. The emergence, zenith and decline of the Greek Republic, Roman Empire, Ming Dynasty in China and European Colonial Empires (including the British, French, Spanish, Portuguese and Dutch) illustrate the opportunities and challenges of riding the waves of innovation. Each gained ascendancy through innovation and each ultimately failed to sustain dominance in the face of new, competing technological or organizational innovations. Currently, the member states of the OECD, led by the US, are arguably at a zenith (or at least local peak) of power and influence. Their future will in part be determined by how they respond to change. If they fail to adapt, other nation states that do may catch up and surpass them in terms of capacity and power. Alternatively, there is some concern that nation states could become irrelevant. UCLA professor Richard Rosecrance has hypothesized that a new 'virtual state' system could evolve, with a network of state and non-state actors engaged in disembodied innovation and production systems, analogous to the hypothesized virtual firm.[2] Whatever the likelihood of either future, the perception in state capitals, in boardrooms and in the minds of many public intellectuals and leaders in civil society is that the risks and opportunities are real and must be acted upon.

## GOVERNING CHALLENGES

British historian Ronald Wright cautions us to remember that the 'idea of material progress is a very recent one – 'significant only in the past three hundred years or so' – coinciding closely with the rise of science and industry and the corresponding decline of traditional beliefs'.[3] Thus, our experience and understanding of change, especially our current exposure to overlapping, interlocking, inexorable technical, economic, social and cultural change, is a relatively recent circumstance. At the same time, many of the institutions we expect to govern the change process are products of an earlier and different period. It may be unrealistic to expect them to be able to respond easily or totally to the new pressures.

It is important to keep in mind that in this relatively recent period of 'progress' we have done reasonably well. There are few glaring examples of significant, irreversible type I errors – that is, where we have either made errors of commission or have been unable to act to block, isolate and withdraw inherently unworthy or dangerous technologies once their adverse effects are known. In almost every case, new technologies have generated net social benefits, with the benefits far outweighing the costs. That is not to say that some technologies in some specific uses haven't caused harm to some (either anticipated or unanticipated) but, rather, is to say that those harms, though significant for individuals, have been relatively small and insignificant in the context of the use of the entire technology over time or space. More often one can point to circumstances where our inability to respond to a technology or innovation has led to it being delayed or abandoned, triggering a type II error, with the concomitant opportunity cost of not acting. One example was the decision to not adopt food irradiation as a cold pasteurization technique to enhance food safety. The resulting instability and waste can be directly linked to states, markets and civil authorities being unable to adapt individually or in unison.

Nevertheless, our prevailing systems are challenged to remain efficient, effective and equitable in three significant ways. First, there is a serious question about the nature of the governing challenge – too little or too much control can be equally damaging to transformative change. Second, there would appear to be a greater degree of path dependence slipping into our economies and societies. Although that offers real hope for some, it presents serious challenges for others. Third, in the post-industrial, knowledge-based world we now live in, individuals and groups increasingly want, and probably need, to feel connected to the process of governing in ways that would have been unimaginable in past years.

## Finding the Right Amount of Governing for Optimal Transformative Innovation

As discussed in Chapter 4, Rosenberg asserts that transformative technology requires 'the freedom to conduct experiments'. Given the uncertainty about where, when and how new transformative technologies can be usefully adapted and adopted (and by extension an inability to identify who should be engaged), there is probably a fine balance between over-governing and under-governing the space for transformative innovation. This poses a serious challenge for societies that wish to make a serious commitment to technological innovation.

The analysis presented in Chapters 9–12 illustrated the multifaceted roles that states, markets and civil authorities can play. States tend to be most important and influential in handling those aspects of the innovation process where there are true public goods, such as in basic research, in rule making (both for markets and for property) and in the evaluation and enforcement of public health and safety. Markets are disproportionately effective whenever the innovation process involves transactions, such as in the commercialization, production and marketing of new technology. Civil authorities are most effective, and one might argue the only credible actors, in defining our knowledge base and for creating collegial or collective rules for the definition of preferences related to goods and services. Although each set of actors has some core areas of competence, they all have increasingly engaged with other sectors to handle those hybrid activities which exhibit a blend of public, private or collective attributes. This area of overlap is perhaps the area at the highest risk of being over- or under-governed.

Innovation, involving the development, gestation, adaptation, adoption and sustained use of new ideas, is likely to flourish only where there is adequate room to test new approaches, appropriate rewards for success and capacity to recover or recycle assets from failures. The tolerance to novel ideas and any corresponding success or failure fundamentally challenges governing systems. Novelty and diversity, though honored in theory, are often looked on unfavorably or actively discouraged by those who hold power. There is a long history in most societies of the tyranny of the majority. In the context of innovation, scholars are suggesting that tolerance of a much larger range of diversity may be vital to attracting, retaining or motivating creative and talented individuals.[4] Furthermore, regardless of any potential long-term benefits of change, there are always likely to be more people and institutions that will prefer the status quo, because they value the immediate or short-term gains more highly than the uncertain prospect of higher long-term benefits. Mancur Olson's observation that it is hard to mobilize people to act positively in their own self-interest when the benefits are widely distributed is germane here. The problem of mobilizing people in support of long-term change is compounded because of our strong preference for short-term results and the corresponding challenge of actually identifying who might have a stake in the change.

A further complication is that there is significant intolerance in many quarters to either major success or catastrophic failures. As incomes and welfare have risen, we have not always become more satisfied and content. There is evidence that consumers and citizens are increasingly influenced by their perceptions of their relative incomes and wealth rather than their absolute status. Extremes of wealth, though admired in the popular press and culture, elicit strong negative reactions from many individuals, states and civil societies. This sense of injustice at times leads governments to rebalance outcomes by recovering 'windfall' gains through taxation, legislation or policy. Similarly, there is significant unease in boardrooms, governments and civil authorities at losses sustained in the innovation process. As effort is increasingly networked in the inventive and gestation phases, more actors perceive they are acting as paymasters or partners in the process. There is a strong preference to win rather than lose. One result has been the rise of what one might characterize as a 'Three Stooges' syndrome, whereby actors who might individually have the capacity to work alone are willing to undertake any risky or uncertain step only with the support (real or perceived) of partners. Thus, there is an increasing tendency for collaboration for the sake of risk management and not simply for knowledge generation. This form of strategy may make sense from an individual organization's perspective – as risks and losses are pooled and discounted and credit for gains are often fully appropriated and counted by each partner – but involves significant costs. In many instances, too many layers of governing rules and structures are involved, with the corresponding loss of creative energy. Similarly, the emergence of transformative change has raised the perception of untenable losses resulting from the 'gales of creative destruction'. Although there may be some areas where the rates of capital investment and innovation are excessive, in most cases the concern is not that the investments are excessive but that the gains are distributed differently from any potential losses. Government monopolies and corporate oligopolies, in particular, are generally unenthusiastic about stranding capital that still has some economic value (such as schools in rural areas, medical equipment that is no longer cutting edge or existing versions of commercial products or services that may be useful but would be abandoned if new versions became available). By the same token, many citizens, consumers and civil authorities demand (if not practice) conservation of capital. One result is that governments and public regulatory processes are under continuous threat of being 'captured' by firms, innovators or civil authorities seeking either to make more room for innovation or to clamp down on the innovation space.

If the resulting governing systems are at root complex, distributed and interconnected networks of institutions that control the creation, evaluation and commercialization of new technologies, then finding how to create space for innovation is challenging. As with any complex system – with multiple levels of interdependent control, numerous subsystems and many subroutines involving a range of public, private and collective actors – there is a possibility that the system may generate small world effects, where small, discrete changes in any one part of the system may have little or no effect or, in some cases, may have profound, unanticipated consequences. Although policy makers and industry could use this in a simplistic way to argue either that changes will have no effect or that any changes would have counterproductive effects, this approach does not need to lead to inaction. Rather, it should lead to a different approach to policy review and development. Thinking of governing as a complex system would suggest that reform should start first with an effort to determine the desired outcomes (including support for innovation, regulatory efficiency, distributional effects or ethical orientation) and then with an examination of the entire complex system as it relates to the desired outcomes. This would in many cases bring in other regulatory and market mechanisms as potential approaches to realize the policy outcome.

One final challenge in governing is that most complex systems ultimately face some diminishing returns. Although making a system more and more complex and interconnected may address problems as they arise, at some point the burden of maintaining such a system is likely to exceed any concurrent benefit. At that point, it often becomes more efficient, effective and equitable to discard the complex system and revert to some local form of governing.[5] Hence, governing systems have to be flexible enough both to expand into areas requiring external management and to retreat when subsidiary capacity has caught up to the challenge.

## Path Dependence

Neoclassical economic theory – involving the concepts of utility and profit optimization and David Ricardo's theory of comparative advantage – leads to the inexorable conclusion that the geographic distribution of production and consumption will be relatively deterministically defined. Theory posits that the working of the invisible hands of a multitude of optimizing actors and agents will align production and trade, such that regions or countries will produce and export goods and services with factor intensities that reflect their underlying relative factor endowments and will correspondingly import goods and services that have factor intensities that do not align with their factor endowments. Hence, countries relatively endowed with land will produce and export goods requiring relatively larger amounts of land (like food or forest products), those relatively endowed with labor will produce goods requiring relatively more workers than other factors (such as textiles) and those relatively endowed with capital will produce and export goods requiring intensive capital investments (such as advanced instrumentation and pharmaceuticals). Although production of a new technology, good or service might conceivably be undertaken in a country without the corresponding comparative advantage, the theory posits that the inexorable forces of the market would drive production toward the country with the appropriate factor endowments. In such a well-defined system, there is little role for the state.[6] Whatever it tries would only have a limited term impact if the neoclassical theory holds.

Princeton political scientist Robert Gilpin argues that several new theories of economic growth potentially create a new role for the state.[7] If the theories are correct, there now is potential for governments to enact policies to change the distribution of production and the rate of economic growth. The new theories of institutional economics,[8] endogenous growth,[9] and strategic trade[10] combine to suggest that there may be localized increasing returns to scale caused by localized external economies of scope (as discussed in Chapter 12). The innovation systems or clusters literature[11] makes an effort to examine what specific features of specific sectors or regions create this kind of geographic lock-in. Most, but not quite all, of this effort is directed to understanding the economic foundations for the apparent path dependency and observed economic agglomeration in sectors such as fashion, culture, biotechnology, computer programming, manufacturing and advanced instrumentation. Although there is undoubtedly economic agglomeration, it is not clear if there truly is any path dependency and, if there is any dependency, what causes it and whether it can be engineered or governed in a proactive way.[12]

One interpretation of the sectoral or technological economic agglomerations we are seeing emerge around the world is that the nature of the innovation process is the driver that generates strong localized, non-traded or location-specific externalities that attract highly innovative firms into clusters. As discussed in Chapter 4, the tacit know-how and know-who knowledge that underpins much research and development related to transformative technologies and the innovation process often involves face-to-face interactions to function. Hence, those areas of innovation involving exploration, creativity and artifice will tend to have a stronger incentive to agglomerate. The economic approach assumes that these linkages and effects exist in absence of any particular institutional or governing system. In contrast, the examination in Chapters 6–12 offers evidence that many of those relationships are embedded in specific, proactive governing subroutines. The efficiency, effectiveness and equitable nature of these systems would appear to be strong determinants of whether innovative activity can occur successfully.

Governing institutions are clearly one significant variable that differs between sectors, technologies and regions. Some transformative technologies are well served by comprehensive, enabling, flexible governing regimes, such that they are able to safely, expeditiously and economically develop and introduce a cascade of innovations. Others face disjointed, inconsistent, cumbersome and expensive governing regimes, with the result that any technological candidates are ill-prepared to enter the market successfully. The evolution of those governing systems is far more likely to exhibit path dependency than arm's-length transactions, if for no other reason than that most institutions are, by definition, a product of historical factors and tend to be located wholly or at least dominantly within a defined region or community.[13] Governments by definition are located in the region that elects and funds them, the vast majority of firms are locally based (even most large multinational enterprises have a majority of their added value, markets and management in their country of origin)[14] and the vast majority of interests and values-based civil authorities are primarily located or connected to a well delineated, geographically connected group of active supporters, adherents and members. Hence, location still does matter, especially for institutions that claim a role in governing technological change.

This point has not gone unnoticed by others. There has been significant empirical research in recent years as people have attempted to determine the scope and impact of the relationship between the nature of governing systems and various economic outcomes. A range of empirical studies have been done: some have looked at correlations between social capital, state governance and economic outcomes;[15] some have looked at the relationship between government efficiency (such as corruption, rule of law, macroeconomic policies and trade policy orientation) and growth;[16] and others have tested whether democracy is a determinant of growth.[17] This study differs from the other work in that it defines governing as more than what governments do. As discussed in earlier chapters, states, markets and civil authorities all engage in governing parts of many systems. The empirical work to date is at best too narrow (if one accepts governments might be a credible and appropriate proxy for the capacity of markets and civil authorities) or at worst mis-specified (if the capacity of markets and civil authorities to govern varies widely from government capacity).

This study would tend to suggest that path dependency will vary depending on the nature of the event being governed. If the event fits in one of the pure domains, where goods are unambiguously public, private or collective, then there may not be much path dependence. The competition between different institutions would drive the system towards the optimal public, private or collective institutional system.[18] So, if a state attempted to deliver a private good, it would ultimately find it pushed out of that domain. The same would be true if a for-profit company tried to produce a collective or public good or if a civil authority tried to produce a pure private or public good. But if a transformative technology involves a good that fits in one of the hybrid areas (where goods exhibit a mix of public, private and collective elements, including rivalry, excludability and voice), then which institution begins will matter a lot.

If state governments take the lead, then they will tend to concentrate the costs and benefits locally, creating the basis for strong path dependency and localization of production. Regardless of the rhetoric, governments remain fundamentally based on unique, non-overlapping geography and socio-economic units. Each international system is at root based on national priorities, national authorities and national enforcement. The almost total absence of any true democracy beyond the nation-state level would appear to create a real and arguably perpetual limit on the internationalization of state government. That said, there are major differences in national and local governing systems. The World Bank, among others, has demonstrated that different jurisdictions have widely varying capacities. As discussed in Chapter 6, the fundamental underpinning of most state-based governing systems is the legal system – common law, civil law, religious law or customary law. Although all of those systems adapt slowly, arguably, the common law system, which places the burden of proof on the state or other actors to prove fault, would seem to be perhaps more accommodating of transformative change than the other systems, because they often need to define proactively new rights and privileges to enable change, which can be difficult or impossible for transformative changes.

If markets lead, they in theory could transcend the state, but in practice they remain highly linked to their geographic and socio-political roots. Markets are for the most part grounded in the legal and social systems of their host states. They fundamentally depend on the rules, norms and conventions which allow individuals to coordinate their transactions via the price mechanism. As such, the environment for markets is determined outside the transaction and for the most part within nation states. This is especially important for transformative technologies, as their higher degree of risk and uncertainty means that full, market-based transactions cannot always be structured through fully specified contracts. Rather, market actors find certainty and security in comfortable, known surroundings where incompletely specified transactions are less risky.

Finally, many civil authorities, especially values-based and democratically organized interest-based groups, aspire to being the voice of the people. The challenge is that many of those organizations reflect a relatively narrow range of values and interests. Even the most international authority is relatively narrowly grounded in terms of members, financial support and targeted messages. Even a cursory review of the LSE Civil Society Database shows that the majority of international civil authorities are largely comprised of northern, developed, Judeo-Christian and capitalist members and led by white, Anglo-European, highly educated males. That doesn't mean that they can't understand and promote a broader range of values and interests but it does suggest that their means, motives and methods are likely disproportionately influenced by a narrow, geographically limited range of perspectives. Similarly, the industrially directed authorities and epistemic communities are highly concentrated in OECD countries and broadly focused on OECD interests.

Which governing structure takes the lead will in the end determine where decisions are made and in whose interest those decisions are made. These choices may ultimately determine the future of any transformative change.

## The ART of Governing

All governing systems, but especially distributed, nested systems of subroutines, involve hierarchies and these hierarchies inevitably create concerns about accountability.[19] As we become familiar with hierarchies, we tend to become more comfortable with their operation. But when faced with transformative change, governing systems are inevitably going to need to adapt and adopt new subroutines and new relationships to manage emerging risks and opportunities. This adaptation process raises three specific challenges: new structures inevitably require new management skills in governments and elsewhere that often are not fully identified or codified in training systems; new structures usually face questions about their accountability; and institutions that aspire to govern on behalf of others, whether old or new, are increasingly challenged to demonstrate their legitimacy.[20] Ultimately, any governing system needs to meet the prevailing standards of good governance, which as discussed in Chapter 6 involve standards of accountability, responsibility and transparency (in essence, the ART of governing).

There is extensive debate and analysis of how ART can be achieved. Although most states have moved to representative forms of government, where those chosen to lead and govern reflect the general interests of those they represent but use their own judgment to make decisions, many forms of delegated governing structures continue in the epistemic communities and industry associations. The ART of governing differs in each circumstance.

In representative systems, operating in most nation-state governments and in many values-based civil authorities, there is often only a weak link between specific interests of the governed and decisions of the governors. In those circumstances, accountability is sustained by a balance of customs, rules and conventions that allow for open, transparent governing processes, procedural links between decisions, outcomes and decision makers and periodic opportunities to assess governors through elections. Although this process can work reasonably well in jurisdictions with clearly delimited elections[21] and full and transparent procedures, it becomes particularly problematic when the power to govern is delegated to bureaucracies or to third parties. In these circumstances, transparency is often lost and questions arise about how responsible and accountable the system really is.

Complex governing systems, in particular, are challenged to remain ARTful. Nested systems involving complex subroutines are always eyed suspiciously by the average citizen, particularly if the procedures are poorly defined and the decisions are difficult to understand in context. Many have called for our governing systems to become more 'reflexive.' At one level, this makes some sense, as the risks and benefits of any transformative technology are usually unknown, and probably unknowable, by any narrow set of governors. A wide number of actors from throughout the commercialization system, ranging from scholars and researchers in the epistemic community to ultimate consumers, can and probably should have a chance to contribute to our assessments of transformative change.

Making our governing systems more reflexive, however, raises two potential problems. First, it will be vitally important for dialogue to be directed to areas where broader input is both valuable and valid. Upstream dialogue which addresses the 'what' of a technology (as in what do we need?, what do we want?, what will we tolerate?) relates directly to values, interests and assumptions and would probably enhance the accountability, responsibility and transparency of our governing systems. The difficulty is that many dialogue processes are poorly structured, so that the debate often focuses on issues of 'how' rather than what. Values, interests and assumptions, though perhaps germane to the earlier stages of review, if engaged at this level often translate into special interests and contribute to regulatory capture. Second, reflexivity is likely not needed in every institution, but rather in the entire system. Thus, we need to find ways to engage in reflexive dialogue both up and downstream in the context of the entire governing system and not in each component of the system.

Finally, reflexivity is perhaps incompatible with the accountable and responsible state and the efficient market. Although only civil authorities can articulate the views of groups in society – both for-profit groups (that self-regulate and manage market functions such as pre-commercial, non-competitive research, standards setting and market development) and not-for-profit social action groups (that represent segments of society that cannot be adequately addressed by private or public actors) – few are truly ARTful. Governments and industry have engaged a wide array of civil authorities in their systems designed to govern transformative technologies but there has been little thought directed to how one can ensure these organizations are truly voicing the views of a segment of society. Many are simply vehicles for special interests. Political scientist Robert Gilpin has suggested that, at the extreme, there is a risk that we may end up with a neo-medieval system where we are over-governed by a complex web of vested interests, analogous to the guild structure of the middle ages. If we wish to avoid such an outcome, the public and private sectors will need to sort out what in their view constitutes a duly constituted civil authority and then work with the various authorities to encourage them to develop appropriate governance models. Although many civil authorities may adequately reflect the view of their members or adherents, the lack of transparency, formal accountability and procedures to ensure responsibility of their leadership undermines their long-term efficacy as vehicles for voice.

In summary, all governing institutions need to aspire much more to be accountable, responsible and transparent in order to create and sustain the legitimacy of the overall governing system and its decisions. Our long-term economic and social future will depend on how we take up this challenge.

## NOTES

1. Maltsbarger and Kalaitzandonakes (2000) modeled the soybean supply chain, developing a proprietary engineering-economic model with hundreds of equations, and Castle *et al.*

(forthcoming) developed an intellectual property/regulatory model that involves an equally large set of relationships.

2. Rosecrance (1996).
3. Wright (2004: 4).
4. For example, Florida (2002).
5. Wright (2004: 92).
6. Irwin (1996) offers an excellent overview of the evolution of our theories of trade and critically examines six of the most common theoretical arguments in favor of government involvement, concluding that none offers much real hope of changing the outcomes anticipated by Ricardo's theory. He thus concurs with Marshall (1890) that free trade is an advantageous expedient because 'the simplicity and naturalness of Free Trade – that is, the absence of any device – may continue to outweigh the series of different small gains which could be obtained by any manipulations of tariffs, however, scientific and astute.'
7. Gilpin (2001), especially Chapters 4 and 5.
8. For example, Coase (1960); Williamson, 1981.
9. For example, Lucas (1988); Romer (1990); Grossman and Helpman (1991).
10. For example, Spencer and Brander (1983); Krugman (1996).
11. Characterized by Lundvall (1992) and Porter (1990).
12. Economists Paul David (1985) and Brian Arthur (1989) developed the concept of technological path dependence.
13. North (1991: 108) makes a similar point.
14. Decoupled, *The Economist*, 23 February 2006.
15. For example, Putnam *et al.* (1993); Putnam (2000).
16. World Bank (1997).
17. For example, Helliwell (1994).
18. In economic terms, one could envisage opportunity costs accumulating with inefficient or ineffective governing systems, such that the system would evolve and new institutions would emerge.
19. Rhodes (1995).
20. Salamon (2002b).
21. Regularly scheduled elections, universal suffrage and direct votes for candidates all contribute to ART. Variably held elections called at the whim of a governing party, limited suffrage and voting for lists of candidates or for delegates who make the final decision loosens the system of accountability and responsibility.

# Bibliography


6, Perri (2001), 'Governing by technique: judgement and prospects for the governance of and with technology', in OECD, *Governance in the 21st Century*, Paris: OECD, accessed 20 August 2005 at www.oecd.org/dataoecd/15/0/17394484.pdf, pp. 67–120.

Acha, V., O. Marsili and R. Nelson (2004), Editorial: 'What do we know about innovation?', *Research Policy*, **33**(9), 1253–8.

AgBios (2005), GM Database, accessed 15 August 2005 at http://agbios.com/dbase. php.

Alston, J., M. Mara, P. Pardey and T. Wyatt (1998), 'Research returns redux: a meta-analysis of the returns to agricultural R&D', *Environment and Production Technology Division discussion paper no. 38*, Washington, DC: International Food Policy Research Institute.

Alston, J., G. Norton and P. Pardey (1995), *Science under Scarcity: Principles and Practice of Agricultural Research Evaluation and Priority Setting*, Ithaca, NY: Cornell University Press.

Amabile, T., R. Conti, H. Coon, J. Lazenby and M. Herron (1996), 'Assessing the work environment for creativity', *The Academy of Management Journal*, **39**(5), 1154–84.

Amaral, L. and J. Ottino (2004), 'Complex networks: augmenting the framework for the study of complex systems', *The European Physical Journal B*, **38**, 147–62.

American Ag-Tec (1998), 'News release: world's first crop of pharmaceutical potatoes to be grown in Wisconsin', accessed at www.quantumtubers.com/pharmaps.htm.

Arndt, H. (1988), 'Market failure and underdevelopment', *World Development*, **16**(2), 219–29.

Arrow, K. (1962), 'The economic implications of learning by doing', *Review of Economic Studies*, **29** (June), 155–73.

Arthur, B. (1989), 'Competing technologies, increasing returns and lock-in by historical events', *Economic Journal*, **99**, 116–31.

Arthur. B. (1994), 'Inductive reasoning and bounded rationality', *American Economic Review*, **84**, 406–11.

Asheim, B. and M. Gertler (2004), 'Regional innovation systems and the geographical foundations of innovation', in J. Fagerberg, D. Mowery, R.

Nelson (eds), *The Oxford Handbook of Innovation*, Oxford: Oxford University Press, pp. 291–317.

Astebro, T. (2003), 'The return to independent invention', *Economic Journal*, **113**, 226–39.

Atkinson, M. and W. Coleman (1992), 'Policy networks, policy communities and the problems of governance', *Governance*, **5**(2), 154–80.

Auditor General of Canada (2002), 'Modernizing accountability in the public sector', Chapter 9 in the *Report of the Auditor General of Canada*, accessed at www.oag-bvg.gc.ca/domino/reports.nsf/html/20021209ce.html.

Audretsch, D. and M. Feldman (1996), 'R&D spillovers and the geography of innovation and production', *American Economic Review*, **96**, 631–40.

Bagehot, W. (1867), *The English Constitution*, reprinted 2001, Cambridge: Cambridge University Press.

Barney, J. (1991), 'Firm resources and sustained competitive advantage', *Journal of Management*, **17**(1), 137–45.

Baty, P. (2005), 'Journals "top ten" sparks a rebellion', *The Times Higher Education Supplement*, 28 January.

Bauer, M. and G. Gaskell (eds) (2002), *Biotechnology: The Making of a Global Controversy*, Cambridge: Cambridge University Press.

Beck, U. (1992), *Risk Society: Towards a New Modernity*, London: Sage.

Benacek, V. (2005), 'Three dimensions of modern social governance: Markets, hierarchies and kinships', Pracovni Sesity working papers, 1/2005.

Benkler, Y. (2004), 'Sharing nicely: on shareable goods and the emergence of sharing as a modality of economic production', *Yale Law Journal*, **114**(2), 273–356.

Benthan, J. (1780), 'Utilitarianism', reprinted 1987 in *Utilitarianism and Other Essays*, London: Penguin Books.

Bentley, A. (1908), *The Process of Government: a Study of Social Pressures*, Chicago: University of Chicago Press.

Bergman, E. and E. Feser (2001), 'Innovation system effects on technological adoption in a regional value chain', *European Planning Studies*, **9**(5), 629–48.

Biemelt, S., U. Sonnewald, P. Galmbacher, L. Willmitzer and M. Müller (2003), 'Production of human papillomavirus type 16 virus-like particles in transgenic plants', *Journal of Virology*, **77**(17), 9211–20.

Bingen, J. and L. Busch (eds) (2005), *Agricultural Standards: The Shape Of The Global Food And Fiber System*, Dordrecht: Springer.

Black, J. (2002), 'Critical reflections on regulation', London School of Economics Centre for Analysis of Risk and Regulation discussion paper 4 accessed 20 August 2005 at http://lse.ac.uk/collections/CARR/pdf/Disspaper4.pdf

Blind, K. (2004), *The Economics of Standards: Theory, Evidence and Policy*, Cheltenham, UK and Northampton, MA, USA: Edward Elgar.

Boulding, K. (ND), 'The limitations of mathematics: an epistemological critique', seminar on the application of mathematics to the social sciences summarized by H. Slater, University of Michigan, 15 December 1955, accessed at http://csf.colorado.edu/boulding/limits_ of_ math.html.

Boulding, K. (1968), *Beyond Economics: Essays on Society, Religion and Ethics*, Ann Arbor, MI: The University of Michigan Press.

Boulding, K. (1970), 'Organizers of social evolution', Chapter 2 in K. Boulding (ed.), *A Primer on Social Dynamics: History as Dialectics and Development*, London: Collier Macmillan, pp.19–36.

Boulding, K. (1984), 'Foreword: a note on information, knowledge and production', in M. Jussawalla and H. Ebenfield (eds), *Communication and Information Economics: New Perspectives*, Amsterdam: North-Holland, pp. vii–ix.

Branstetter, L. and Y. Ogura (2005), 'Is academic science driving a surge in industrial innovation? Evidence from patent citations', National Bureau of Economic Research working paper no. 11561, accessed 3 May 2006 at www.nber.org/papers/w11561.

Bresnahan, T.F. and M. Trajtenberg (1995), 'General purpose technologies "Engines of Growth" ', *NBER working paper* no. 4148.

Brooks, H. (1994), 'The relationship between science and technology', *Research Policy*, **23**, 477–86.

Brown, J. and P. Duguid (2000), *The Social Life of Information*, Boston: Harvard Business School Press.

Bruland, K. and D. Mowery (2005), 'Innovation through time', in J. Fagerberg, D.C. Mowery and R.R. Nelson (eds), *The Oxford Handbook of Innovation*, Oxford: Oxford University Press.

Brunk, C., L. Haworth and B. Lee (1991), *Value Assumptions in Risk Assessment: a Case Study of the Alachlor Controversy*, Waterloo, ON: Wilfred Laurier University Press.

Buchanan, J. (1989), 'The public choice perspective', in J. Buchanan (ed.), *Essays on Political Economy*, Honolulu: University of Hawaii Press, pp. 13–24.

Buckwell, A., G. Brookes and D. Bradley (1999), 'Economics of identity preservation for genetically modified crops', report prepared for Food Biotechnology Communications Initiative.

Buenz, E., D. Schnepple, B. Bauer, P. Elkin, J. Riddle and T. Motley (2004), 'Techniques: bioprospecting historical herbal texts by hunting for new leads in old tomes', *Trends in Pharmacological Sciences*, **25**(9), 494–8.

Burt, R. (2005), *Brokerage and Closure: An Introduction to Social Capital*, Oxford: Oxford University Press, accessed at http://gsbwww.uchicago.edu/fac/ronald.burt /research.

Bush, V. (1945), *Science, the Endless Frontier: A Report to the President on a*

*Program for Postwar Scientific Research*, Washingston DC: United States Office of Scientific Research and Development.

Canadian Biotechnology Advisory Committee (CBAC) (2002), 'Improving the regulation of genetically modified foods and other novel foods in Canada: a report to the Government of Canada Biotechnology Ministerial Coordinating Committee (final report)', Government of Canada, accessed 10 May, 2006, at http://cbac-cccb.ca/epic/internet/incbac-cccb.nsf/vwapj/cbac_report_e.pdf/$FILE/ cbac_report_e.pdf

Carson, R. (1964), *Silent Spring*, Greenwich, CT: Fawcett Publications.

Castells, M. (1996), *The Rise of the Network Society*, Oxford: Blackwell Publishers.

Castle, D. (forthcoming), *Plant Derived Vaccines: Innovations and Regulatory Burdens*, Chichester: John Wiley and Sons.

Caulfield, T. (2002), 'Science, with a bang: the hype unleashed in the rush to cash in on the genetic revolution threatens to doom important research', *Ottawa Citizen*, A15.

Central Intelligence Agency (CIA) (2005), *CIA Factbook*, accessed 5 October, at www.cia.gov/cia/publications/factbook/index.html.

Charbonneau, L. (2005), 'Ethics boards harming survey research', *University Affairs*, June–July, accessed at www.universityaffairs.ca.

Chartrand, H. (2006), 'The competitiveness of nations in a global knowledge-based economy', PhD dissertation for the University of Saskatchewan, Canada.

Christensen, C. (1997), *The Innovator's Dilemma*, Boston, MA: Harvard Business School Press.

Coase, R. (1937), 'The nature of the firm', *Economica*, **4**(16), 386–405.

Coase, R. (1960), 'The problem of social cost', *Journal of Law and Economics*, **3**, 1–44.

Codex Alimentarius (2000), *Origins of the Codex Alimentarius*, accessed 12 October, at www.fao.org/docrep/w9114e/W9114e03.htm.

Commission for Environmental Cooperation of North America (2004), 'Maize and biodiversity: the effects of transgenic maize in Mexico', secretariat article 13 report, 31 August.

Commission of the European Communities (EC) (2000), Communication from the Commission on the Precautionary Principle, COM(2000) 1 final, 26 January, Brussells.

Commission of the European Communities (EC) (2001), 'European governance: a white paper', COM(2001) 428 final, 25 July, Brussels.

Commission on Global Governance (1995), *Our Global Neighbourhood*, Oxford: Oxford University Press.

Committee for Study of Invention (2004), 'Invention: enhancing inventiveness for quality of life, competitiveness, and sustainability', report sponsored by the Lemelson-MIT Program and the National Science Foundation, 23 April, accessed 10 August, 2005 at http://web.mit.edu/invent/n-pressreleases/downloads/report.pdf.

Cooke, P. (2002), *Knowledge Economies: Clusters, Learning and Co-Operative Advantage*, New York: Routledge.

Cooke, R. (1990), *Subjective Probability and Expert Opinion*, Oxford: Oxford University Press.

Council for Biotechnology Information (CBI) (2004), 'A biotech potato holds the clues to medically useful proteins', accessed 15 August, 2005 at http://whybiotech.ca/canada-english.asp?id=4102.

Cranor, C. (no date), 'The normative nature of risk assessment: features and possibilities. RISK: Health, Safety & Environment', accessed at www.piercelaw.edu/risk/vol8/spring/cranor.htm.

Dahl, R. (1998), *On Democracy*, New Haven, CT: Yale University Press.

Dahlman, C. (1979), 'On the problem of externality', *Journal of Law and Economics*, **22**, 141–62.

Dalrymple, D. (2006), 'Impure public goods and agricultural research: toward a blend of theory and practice', *Quarterly Journal of International Agriculture*, **45**(1), 71–89.

Darby, M. and L. Zucker (2003), 'Grilichesian breakthroughs: inventions of methods of inventing and firm entry in nanotechnology', National Bureau of Economic Research working paper 9825, accessed 3 May, 2006 at www.nber.org/papers/w9825

Darwin, C. (1859), *The Origin of Species by Means of Natural Selection: or, the Preservation of Favoured Races in the Struggle for Life* (reprinted 1906), London: J. Murray.

David, P. (1985), 'Clio and the economics of QWERTY', American Economic Review Papers and Proceedings, **75**, 332–37.

Davies, L. (2002), 'Technical cooperation and the international coordination of biotechnology inventions', *Journal of Law and Society*, **29**(1), 137–62.

De Alcantara, C. (1998), 'Uses and abuses of the concept of governance', *International Social Science Journal*, **50**(155), 105–13.

Dear, P. (2004), 'Mysteries of state, mysteries of nature: authority, knowledge and expertise in the 17th century', in S. Jasanoff (ed.), *States of Knowledge: The Co-production of Science and Social Order*, London: Routledge, pp. 206–24.

Diamond, J. (1997), *Guns, Germs, and Steel: the Fates of Human Societies*, New York: W.W. Norton & Co.

Dierker, D. and P. Phillips (2001), 'The search for the holy grail? Freedom to operate in Canadian agricultural biotechnology', proceedings of the International Consortium on Agricultural Biotechnology Research (ICABR) meetings, Ravello, Italy, June.

Dierker, D. and P. Phillips (2002), 'The butcher, the baker, the candlestick maker: why agricultural biotechnology industry may differ from the general biotechnology industry', selected paper at the American Agricultural Economics Association meetings, 28–31 July.

Dierker, D. and P. Phillips (2003), 'Agricultural biotechnology letters of marque', invited paper session 10, AAEA/CAES Meetings, Montreal, 24–27 July.

Doern, B. and T. Reed (2000), *Risky Business*, Toronto: University of Toronto Press.

Dosi, G. (1984), *Technical Change and Industrial Transformation: The Theory and an Application to the Semiconductor Industry*, London: The Macmillan Press.

Dosi, G. and R. Nelson (1994), 'An introduction to evolutionary theories in economies', *Journal of Evolutionary Economics*, **4**(1), 153–72.

Downs, A. (1957), *An Economic Theory of Democracy*, New York: Harper and Brothers.

Dror, Y. (2004), 'The ultimate challenge: helping countries to weave their futures', outline of keynote address UNDP public administration reform practice meeting Bratislava, 21 April, accessed 1 September, 2005 at www.ecissurf.org/files/pubadmin /Dror_Presentation%20Outline.doc.

*The Economist* (2004), 'Passing on the crown', **373**(8400), 69–71.

*The Economist* (2005), 'Eminent domain', 9 July, 56.

Eggertsson, T. (1995), 'Economic perspectives on property rights and the economics of institutions', in P. Foss (ed.), *Economic Approaches to Organizations and Institutions: An Introduction*, Dartmouth, UK: Dartmouth Publishing Co.

Einsiedel, E., K. Finlay and J. Arko (2000), 'Meeting the public's need for information on biotechnology', commissioned paper for Canadian Biotechnology Advisory Committee, accessed 26 June, 2002 at www.cbac-cccb.ca/documents/en/PublicNeed_Info_Einsiedel.pdf.

Elkins, D. (1995), *Beyond Sovereignty*, Toronto: The University of Toronto Press.

Environics International (2000), 'Biotechnology for medicine and cloning: yes or no?', accessed at www.globescan.com/news_archives/IEM_biotechnology.pdf.

Etzkowitz, H. and L. Leydesdorff (1995), 'The triple helix: university-industry-government relations. A laboratory for knowledge based economic development', *European Society for the Study of Science and Technology Review* **14**, 14–19.

Etzkowitz, H. and L. Leydesdorff (2000), 'The dynamics of innovation: from national systems and 'Mode 2' to a triple helix of university-industry-government relations', *Research Policy*, **29**(2), 109–23.

Evenson, R. and D. Gollin (2003), 'Assessing the impact of the green revolution, 1960 to 2000', *Science* **300**(5620), 758–62.

Evenson, R. and V. Santaniello (2004), *Consumer Acceptance of Genetically Modified Foods*, Oxford: CABI.

Ewens, S. and A. Pusztai (1999), 'Health risks of genetically modified foods', *The Lancet* **354**, 684.

Fagerberg, J., D. Mowery and R. Nelson (eds) (2005), *The Oxford Handbook of Innovation*, Oxford: Oxford University Press.

Florida, R. (2002), *The Rise of the Creative Class and How It's Transforming Work, Leisure and Everyday Life*, New York: Basic Books.

Frankl, V. (1962), *Man's Search for Meaning*, Boston, MA: Beacon Press.

Freeman, C. (1987), *Technology Policy and Economic Performance: Lessons from Japan*, London/New York: SPRU/Pinter Publishers.

Freeman, C. (1988), 'Japan: a new national system of innovation?', in G. Dosi, C. Freeman, G. Silverberg and L. Soete (eds), *Technical Change and Economic Theory*, London and New York: Pinter Publishers, pp. 330–348.

Friederichs III, N.P. (1998), 'Patent strategies', *Patent Café Magazine*, accessed at www.patentcafe.com/café_magazine/981115-fl.html.

Fukyama, F. (2002), *Our Posthuman Future: Consequences of the Biotechnical Revolution*, New York: Farrar, Straus and Giroux.

Gaskell, G., E. Einsiedel, S. Priest, W, Hallman, and J. Olsthoorn (2005), 'Social values and the governance of science', *Science*, **310**(5756), 1908–9.

Gibbons, M., C., Limoges, H. Nowotny, S. Schwartzman, P. Scott and M. Trow (1994), *The New Production of Knowledge: The Dynamics of Science and Research in Contemporary Societies*, London: Sage.

Giddens, A. (1994), 'Living in a post-traditional society', in U. Beck, A. Giddens and S. Lash (eds), *Reflexive Modernity*, Cambridge: Polity.

Giddens, A. (1997), *Sociology*, 3rd edn, Cambridge: Polity Press.

Giddens, A. (1998), *The Third Way: The Renewal of Social Democracy*, Cambridge: Polity Press.

Giles, J. (2005), 'Internet encyclopedias go head to head', *Nature*, **438** (15 December), 900–901.

Gilipin, R. (1991), *Global Political Economy: Understanding the International Economic Order*, Princeton, NJ: Princeton University Press.

Gold, R. (2001), 'Patenting life forms: an international comparison', study prepared for the Canadian Biotechnology Advisory Committee, accessed 10 May, 2006 at http://cbac-cccb.ca/epic/internet/incbac-cccb.nsf/en/ah00399e.html#executive.

Goodman, S., J. Berlin, S. Fletcher and R. Fletcher (1998) 'Manuscript quality before and after reviewing and editing at *Annals of Internal Medicine*', *Annals of International Medicine*, **121**, 11–21.

Grant, R. (2002), 'The knowledge based view of the firm', in C. Choo and N.

Bontis (eds), *The Strategic Management of Intellectual Capital and Organizational Knowledge*, New York: Oxford University Press.

Gray, R., S. Malla and P. Phillips (2006), 'Product innovation in the Canadian canola sector', *Supply Chain Management: An International Journal*, **11**(1), 65–74.

Grossman, G. and E. Helpman (1991), *Innovation and Growth in the Global Economy*, London: The MIT Press.

Gunningham, N. and P. Grabosky (1999), *Smart Regulation: Designing Environmental Policy*, New York: Oxford University Press.

Haas, P. (1992), 'Introduction: epistemic communities and international policy coordination', *International Organization*, **46**(1), 1–35.

Hall, B. (2004), 'Innovation and diffusion', National Bureau of Economic Research working paper 10212, accessed 3 May, 2006 at www.nber.org/papers/w10212.

Handelman, S. (1994), 'The Russian Mafiya', *Foreign Affairs* **73**(2), 83–96.

Hardin, R. (1999), 'Street-level epistemology and democratic participation', proceedings of the European Public Choice Society, Lisbon, 7–10 April, accessed 25 October at http://cniss.wustl.edu/rikerpapers/hardinpaper1.html

Harl, N.E., R.G. Ginder, C.R. Hurburgh and S. Moline (2000), 'The Starlink™ situation', accessed at http://www.extension.iastate.edu/Pages/grain/publications/buspub/ 0010star.PDF.

Hawksworth, D. (1992), 'The need for a more effective biological nomenclature for the 21st century', *Biological Journal of the Linnean Society*, **109**, 543–67.

Hawksworth, D. (1995), 'Steps along the road to a harmonized bionomclature', *Taxon* **44**, 447–56.

Held, D., A. McGrew, D. Goldblatt and J. Perraton (1999), *Global Transformations: Politics, Economics and Culture*, London: Polity.

Helliwell, J. (1994), 'Empirical linkages between democracy and economic growth', *British Journal of Political Science*, **24**, 225–48.

Helpman, E. (ed.), (1998), *General Purpose Technologies and Economic Growth*, Cambridge, MA: MIT Press.

Hepworth, L. (2001), 'Industry stewardship as a response to food safety concerns', Chapter 5 in P. Phillips and R. Wolfe (eds), *Governing Food: Science, Safety and Trade*, Montreal: McGill University Press/Queens School of Policy Studies.

Hicks, J. (1932), *The Theory of Wages*, London: Macmillan and Co., Ltd.

Hirschman, A. (1970), *Exit, Voice and Loyalty*, Cambridge, MA: Harvard University Press.

Hobbes, T. (1651), *The Leviathan*, reprinted 1998, Oxford: Oxford University Press.

Homer-Dixon, T. (2001), *The Ingenuity Gap*, Toronto: Vintage Canada.

Hughes, T. (1989), *American Genius: A Century of Invention and Technological Enthusiasm 1870–1970*, New York: Viking.

Interactive European Network for Industrial Crops and their Applications (IENICA) (ND), website accessed 15 August, 2005 at www.ienica.net/crops/potatoes .htm

International Association for Public Participation (2000), 'IAP2 Public Participation Spectrum', accessed 30 August, 2005 at www.iap2.org/associations/4748 /files/spectrum.pdf

Internet Systems Consortium, Inc. (2005), 'ISC internet domain survey', accessed 30 August, 2005 at www.isc.org/index.pl?/ops/ds/.

Irwin, D. (1996), *Against the Tide: An Intellectual History of Free Trade*, Princeton, NJ: Princeton University Press.

Isaac, G. (2002), *Agricultural Biotechnology and Transatlantic Trade: Regulatory Barriers to GM Crops*, Oxford: CABI Publishing.

James, C. (Various), 'Global status of commercialized transgenic crops', International Service for the Acquisition of AgrioBiotech Applications briefs, accessed at www.isaaa.org

Jasanoff, S. (1987), 'Contested boundaries in policy-relevant science', *Social Studies of Science*, **17**(2), 195–230.

Jasanoff, S. (ed.) (2004), *States of Knowledge: The Co-production of Science and Social Order*, London: Routledge.

Jensen, M.B., B. Johnson, N. Lorenz and B.-A. Lundvall (2004), 'Codification and modes of innovation', paper presented at the DRUID Summer Conference, Elsinore, Denmark, 14–16 June.

Jessop, B. (1998), 'The rise of governance and the risks of failure: the case of economic development', *International Social Science Journal*, **50**(155), 20–45.

Jevons, W. (1871), *Principles of Science*, reprinted 2003, Honolulu, HI: University Press of the Pacific.

Jones, B. (2004), 'The burden of knowledge and the 'death of the renaissance man': Is innovation getting harder?', National Bureau of Economic Research working paper 11360, accessed 10 August, 2005 at www.nber.org/papers/w11360.pdf.

Joskow, P. (1987), 'Contract duration and relationship-specific investments: empirical evidence from coal market', *American Economic Review*, **71**, 168–85.

Josling, T., D. Roberts and D. Orden (2002), *Food Regulations and Trade: Toward a Safe and Open Global System*, Washington, DC: Institute for International Economics.

Kahn, R. (1994), 'The role of government in the evolution of the internet', in National Academy of Engineering (ed), *Revolution in the US Information*

*Structure*, accessed on 18 August, 2005 at www.nap.edu/readingroom/books/newpath/chap2.html.

Kalaitzanondakes, N. (ed.) (2003), *The Economic and Environmental Impacts of Agbiotech: A Global Perspective*, New York: Kluwer Academic.

Kaldor, M., H. Anheier and M. Glasius (eds) (2003), *Global Civil Society 2003*, Oxford: Oxford University Press, accessed 20 August, 2005 at http://lse.ac.uk/Depts/global/yearbook03chapters.htm.

Kaldor, M., H. Anheier and M. Glasius (eds) (2004), *Global Civil Society 2004/5*, London: Sage, accessed 20 August, 2005 at http://lse.ac.uk/Depts/global/yearbook04chapters.htm.

Kalthoff, O., I. Nonaka and P. Nueno (1997), *The Light and the Shadow: How Breakthrough Innovation is Shaping European Business*, Oxford: Capstone Publishing.

Kaufmann, D., A. Kraay and M. Mastruzzi (2004), 'Governance Matters III', *The World Bank Economic Review*, **18**(2), 253–87.

Kazancigil, A. (1998), 'Governance and science: market-like modes of managing society and producing knowledge', *International Social Science Journal*, **50**(155), 69–79.

Kennett, J., M. Fulton, P. Molder and H. Brooks (1998), 'Supply chain management: the case of a UK baker preserving the identity of Canadian milling wheat', *Supply Chain Management* **3**(3), 157–66.

Kieff, S. (ed.) (2003), *Perspectives on Properties of the Human Genome Project*, Amsterdam: Academic Press.

Kline, S. and N. Rosenberg (1986), 'An overview of innovation', in R. Landau, and N. Rosenberg (eds), *The Positive Sum Strategy: Harnessing Technology for Economic Growth*, Washington, DC: National Academy Press.

Knoke, D. and J. Kuklinski (1982), 'Network analysis', Sage University Quantitative Applications in the Social Sciences series, paper 28.

Krugman, P. (ed.) (1996), *Strategic Trade Policy and the New International Economy*, Cambridge, MA: MIT Press.

Krugman, P. (1998), 'What's new about the new economic geography?', *Oxford Review of Economic Policy*, **14**(2), 7–17.

Kuhn, T. (1970), *The Structure of Scientific Revolutions, 2nd edn*, Chicago, IL: University of Chicago Press.

Laestadius, S. (2000), 'Biotechnology and the potential for a radical shift of technology in forest industry', *Technology Analysis and Strategic Management*, **12**(2), 193–212.

Landes, D. (1999), *The Wealth and Poverty of Nations: Why Some are so Rich and Some so Poor*, New York: WW Norton and Co.

Leifer, R., C. McDermott, G. O'Connor, L. Peters, M. Rice, and R. Veryzer (2000), *Radical Innovation: How Mature Companies can Outsmart Upstarts*, Cambridge, MA: Harvard Business School Press.

Lemelson-MIT Program (1997), 'Inventor of the week: cloning of genetically engineered molecules', accessed 1 September, 2005 at http://web.mit.edu/invent/iow/boyercohen html.

Leydesdorff, L. (2005), 'The knowledge-based economy and the triple helix', in W. Dolfsma and L. Soete (eds), *Reading the Dynamics of a Knowledge Economy*, Cheltenham, UK and Northampton, MA, USA: Edward Elgar.

Lezaun, J. and Y. Millo (2005), 'Regulatory experiments: putting GM crops and financial markets on trial', *Economic and Social Research Council (ESRC) Centre for the Analysis of Risk and Revolution (CARR) discussion paper 30*, London School of Economics, February, accessed 20 August, 2005 at http://lse.ac.uk/collections/CARR/pdf/Disspaper30.pdf.

Lin, W. (2002), 'Segregation of non-biotech corn and soybeans: who bears the cost?', paper presented to the 6th International Conference of the International Consortium on Agricultural Biotechnology Research, Ravello, Italy, 11–14 July.

Lipsey, R. (1995), 'Policy challenges in the new economy', *Micro*, **10**(1), 2.

Lipsey, R. and C. Bekar (1995), 'A structuralist view of technical change and economic growth', *Bell Canada Papers on Economics and Policy 3, Proceedings of the Bell Canada Conference at Queen's University, November 1994*, Kingston, Ont: John Deutsch Institute for the Study of Economic Policy.

Locke, J. (1689), *A letter concerning toleration*, reprinted 1983, Cambridge, MA: Hackett Publishing.

Loder, N. (2005), 'Small wonders: a survey of nanotechnology', *The Economist*, Survey Section, **374**(8407), 3–4.

Losey, J., L. Raynor and M. Carter (1999), 'Transgenic pollen harms Monarch larvae', *Nature*, **399**(6733), 214.

Lucas, R. (1988), 'On the mechanics of economic development', *Journal of Monetary Economics*, **22**, 30–42.

Lundvall, B.-Å. (1988), 'Innovation as an interactive process: from user-producer interaction to the national system of innovation', in G. Dosi, C. Freeman, G. Silverberg and L. Soete (eds), *Technical Change and Economic Theory*, London: Pinter Publishers, pp. 349–69.

Lundvall. B.-Å. (1992), *National Systems of Innovation: Towards a Theory of Innovation and Interactive Learning*, New York: Pinter.

Lundvall, B.-Å. and B. Johnson (1994), 'The learning economy', *Journal of Industry Studies*, **1**(2), 23–42.

Lux Research Corp. (2005), homepage of Lux Nanotech Index™, accessed 11 August, 2005 at http://www.Luxresearchinc .com/luxni.html.

Lynn, L., C. Heinrich and C. Hill (2000), 'Studying governance and public management: Why? How?', in C. Heinrich and L. Lynn (eds), *Governance*

*and Performance: New Perspectives*, Washington, DC: Georgetown University Press.
Machiavelli, N. (1513), *The Prince*, reprinted 2003, London: Penguin Books.
Mahoney, J. (1992), 'The choice of organisational form: vertical financial ownership versus other methods of vertical integration', *Strategic Management Journal* **13**, 559–84.
Maillat, D. (1991), 'The innovation process and the role of the milieu', in E. Bergman, G. Maier and F. Tootling (eds), *Regions Reconsidered: Economic Networks, Information, and Local Development*', London: Mansell, pp. 103–17.
Majone, G. (1994), 'The rise of the regulatory state in Europe', *West European Politics*, **17**(3), 77–102.
Majone, G. (2000), 'The credibility crisis of community regulation', *Journal of Common Market Studies*, **38**(2), 273–302.
Malbera, F. and L. Orsenigo (1990), 'Technological regimes and patterns of innovation: a theoretical and empirical investigation of the Italian case', in A. Heertje and M. Perlman (eds), *Evolving Technology and Market Structure*, Ann Arbor, MI: University of Michigan Press.
Malecki, E. (1997), *Technology and Economic Development: the Dynamics of Local, Regional and National Competitiveness*, Toronto: Longman.
Maloney, B., N. Takeda, Y. Suzaki, Y. Ami, T. Li, T. Mivamura, C. Arntzen and H. Mason (2005), 'Challenges in creating a vaccine to prevent hepatitis E', *Vaccine* **23**(15), 1870–4.
Maltsbarger, R. and N. Kalaitzandonakes (2000), 'The hidden costs of identity preservation', *Feedstuffs*, **72**(36), 1–2.
Marshall, A. (1890), *Principles of Economics*, London: Macmillan.
Maslow, A. (1943), 'A theory of human motivation', *Psychological Review* **50**, 370–96.
Marks, L., and N. Kalaitzandonakes (2001), 'Mass media communications about agrobiotechnology', *AgBioForum*, **4**(3&4), 199–208, accessed at www.agbioforum.org.
Maskell, P. and M. Lorenzen (2003), 'The cluster as market organization', DRUID Working Papers 03–14, DRUID, Copenhagen Business School.
Mason, H., J. Ball, J.J. Shi, X. Jiang, M. Estes and C. Arntzen (1996), 'Expression of Norwalk virus capsid protein in transgenic tobacco and potato and its oral immunogenicity in mice', *Proceedings of the National Academy of Science* **93**, 5335–40.
McKitrick, R. (forthcoming), 'Bringing balance, disclosure and due diligence into science-based policy making', in J. Porter and P. Phillips (eds), *Public Science in Liberal Democracy*, Toronto: University Toronto Press.
McLeod, A. (ed.) (1979), *The Story of Rapeseed in Western Canada*, Regina, Saskatchewan Wheat-Pool.

McNeill, J. (1996), 'The BioCode: integrated biological nomenclature for the 21st century?', in J. Reveal (ed.), proceedings of a mini-symposium on biological nomenclature in the 21st century, held at the University of Maryland, 4 November.

Mead, S. (2001), 'Using social network analysis to visualize project teams', *Project Management Journal*, **32**(4), 32–8.

Mehta, M.D. (2003), 'Regulating biotechnology and nanotechnology in Canada: a post-normal science approach for inclusion of the fourth helix', in Z. Baber and H. Khondker (eds), *The Triple Helix*, Albany, NY: State University of New York Press.

Mehta, M. (2006), 'Paradigm waas: a post-normal science approach to regulating biotechnology and nontechnology', in Phillips et al. (2006).

Merton, R. (1942), 'The normative structure of science', reprinted in N. Storer (ed.) (1973), *The Sociology of Science: Theoretical and Empirical Investigations*, London: University of Chicago Press, pp. 267–78.

Metcalfe, S. (1995), 'Technology systems and technology policy in an evolutionary framework', *Cambridge Journal of Economics*, **19**, 25–46.

Michalski, W., R. Miller and B. Stevens (2001), 'Governance in the 21st century: power in the global knowledge economy and society', in *Governance in the 21st Century*, Paris: OECD.

Mill, J. (1848), *Principles of Political Economy and Chapters on Socialism*, reprinted 1998, Oxford: Oxford University Press.

Mills, L. (2002), *Science and Social Context: The Regulation of Recombinant Bovine Growth Hormone in North America*, Montreal: McGill-Queen's University Press.

Mokyr, J. (1997), 'The political economy of technological change: resistance and innovation in economic history', accessed 15 September, 2005 at www.faculty.econ.northwestern.edu/faculty/mokyr/berg.pdf.

Mokyr, J. (2002a), 'Innovation in an historical perspective: tales of technology and evolution', in B. Steil, G. Victor and R. Nelson (eds), *Technological Innovation and Economic Performance*, Princeton, NJ: Princeton University Press.

Mokyr, J. (2002b), *The Gifts of Athena: Historical Origins of the Knowledge Economy*, Princeton, NJ: Princeton University Press.

Murray, A. (2003), 'Regulation and rights in networked space', *Journal of Law and Society*, **30**(2), 187–216.

Medical Research Council (MRC) (2005), 'Good research proactive', MRC ethics series, accessed 7 February, 2006 at www.mrc.ac.uk/pdf-good-research-practice.pdf.

National Academy of Sciences Committee on the Institutional Means for Assessment of Risks to Public Health, Commission of Life Sciences (1983), *Risk Assessment in the Federal Government: Managing the Process*, Washington, DC: National Academy Press.

National Research Council (1992), *From Rapeseed to Canola: The Billion Dollar Success Story*, Saskatoon: National Research Council.

National Venture Capital Association (2004), 'Venture impact 2004: venture capital benefits to the US economy', accessed 10 May, 2006, at www.nvca.org/pdf/VentureImpact2004.pdf.

Nelson, R. (1995), 'Recent evolutionary theorizing about economic change', *Journal of Economic Literature*, **33**(March), 48–90.

Nesta, L. and V. Mangematin (2004), 'Dynamics of innovation networks', SPRU electronic working paper series 114, April.

Nightingale, P. (1998), 'A cognitive model of innovation', *Research Policy*, **27**, 689–709.

North, D. (1991), 'Institutions', *Journal of Economic Perspectives*, **5**, 97–112.

North, D. (1993), 'Economic performance through time', Nobel Prize lecture, 9 December, accessed on 31 May 2006 at http://nobelprize.org/economics/laureates/1993/north-lecture.html.

North, D. (2005), *Understanding the Process of Economic Change*, Princeton, NJ: Princeton University Press.

Novos (2006), 'Obtaining a business method patent', accessed at www.nolo.com/article.cfm/objected/C2DBFF26–7097–4878–AE36DA00499851EE/11177/167/ART/.

OECD (1992), *Technology and the Economy: The Key Relationships*, Paris: OECD.

OECD (1995), *Interpreting Technological Innovation Data (Oslo Manual)*, 2nd edn, Paris: DSTI/OECD/European Commission Eurostat, 31 December.

OECD (1996), *The Knowledge Based Economy*, Paris: OECD, accessed at www. oecd.org/dsti/sti/s_t/inte/prod/kbe.htm.

OECD (1997), *The Measurement of Scientific and Technical Activities: Proposed Guidelines for Collecting and Interpreting Technological Innovation Data (Oslo Manual)*, Paris: OECD, accessed 2 September, 2005 at www1.oecd.org/dsti/sti/stat-ana/prod/oslo-eng.pdf.

OECD (1999), *Policy Brief*, Paris: OECD, accessed at www.oecd.org.

OECD (2000), *The OECD Guidelines for Multinational Enterprises, Revision 2000*, Paris: OECD.

OECD (2001), *Governance in the 21st Century*, Paris: OECD, accessed 20 August, 2005 at www.oecd.org/dataoecd/15/0/17394484.pdf.

OECD (2005a), *Governance of Innovation Systems, Volume 1: Synthesis Report*, Paris: OECD.

OECD (2005b), *OECD Science, Technology and Industry Scoreboard 2005: Towards a Knowledge-based Economy*, Paris: OECD.

OECD (2005c), 'An introduction to the biosafety consensus documents of OECD's working group for harmonization in biotechnology, series on

Harmonization of Regulatory Oversight in Biotechnology no. 32, ENV/JM?MOND(2005)5, 22 February, Paris.

Olson, M. (1965), *The Logic of Collective Action: Public Goods and the Theory of Groups*, Cambridge, MA: Harvard University Press.

Olson, M. (1982), *The Rise and Decline of Nations: Economic Growth, Stagflation, and Social Rigidities*, New Haven, CT: Yale University Press.

Ogilvy, J. (1986–7), 'Scenarios for the future of governance', *The Bureaucrat*, 13–16.

Ortiz-Garcia, S., E. Ezcurra, B. Schoel, F. Acevedo, J. Soberon and A. Snow (2005), 'Absence of detectable transgenese in local landraces of maize in Oaxaca, Mexico (2003–2004)', *PNAS Early Edition*, accessed at www.pnas.org/cgi/doi/10.1073/pnas.0503356102, pp 1–6.

Osborne, D. and T. Gaebler (1992), *Reinventing Government: How the Entrepreneurial Spirit is Transforming the Public Sector*, London: Plume/Penguin.

Pagden, A. (1998), 'The genesis of "governance" and Enlightenment conceptions of the cosmopolitan world order', *International Social Science Journal*, **50**(155), 7–15.

Paquet, G. (2001), 'The new governance, subsidiarity and the strategic state', in Organisation for Economic Co-operation and Development, (OECD), *Governance in the 21st Century*, Paris: OECD, accessed 20 August, 2005 at www.oecd.org/dataoecd/15/0/17394484.pdf.

Pal, L. and J. Maxwell (2003), 'Assessing the public interest in the 21st century: a framework', paper prepared for the External Advisory Group on Smart Regulation (Canada), chaired by Gaetan Lussier, December.

Pavitt, K. (2003), 'The process of innovaton', SPRU Electronic working paper series 89, August, accessed 10 August, 2005 at www.sussex.ac.uk/Units/spru/publications/imprint/sewps/sewp89/sewp89.pdf.

Penrose, E. (1959), *The Theory of Growth of the Firm*, Oxford: Blackwell.

Perkins, D. (2004), *The Architecture of Invention*, Cambridge, MA: Lemelson-MIT Program.

Persaud, A. (2001), 'The knowledge gap', *Foreign Affairs*, **80**(2), 107–17.

Phillips, P. (2000), 'Intellectual property rights and public research in Canada', in V. Santaniello, R.E. Evanson, D. Zilberman and G.A. Carlson (eds), *Agriculture and Intellectual Property Rights: Economic Institutional and Implementation Issues in Biotechnology*,Wallingford, UK: CAB Publishing.

Phillips, P. (2002), 'Regional systems of innovation as a modern R&D entrepot: the case of the Saskatoon biotechnology cluster', in J. Chrisman, A.D. Holbrook and J.H. Chua (eds), *Innovation, Entrepreneurship, Family Business and Economic Development: A Western Canadian Perspective*, Calgary: University of Calgary Press, pp. 31–58.

Phillips, P. (2003), 'The challenge of creating, protecting and exploiting networked knowledge', proceedings of the International Consortium on Agricultural Biotechnology Research (ICABR) Conference, Ravello, Italy, 29 June–4 July.

Phillips, P. (2005), 'The challenge of creating, protecting and exploiting networked knowledge', in E. Einsiedel and F. Timmermans (eds), *Crossing Over: Genomics in the Public Arena*, Calgary: University of Calgary Press, pp. 7–32.

Phillips, P. and D. Corkindale (2003), 'Marketing GM foods: the way forward', *AgBioForum* **5**(3), 113–21.

Phillips, P. and J. Gustafson (2000), 'Patent strategies in the biotechnology industry and implications for technology diffusion', Proceedings of the International Consortium on Agricultural Biotechnology Research (ICABR) Conference, University of Rome 'Tor Vergata', August, Ravello, Italy.

Phillips, P. and G. Khachatourians (2001), *The Biotechnology Revolution in Global Agriculture: Invention, Innovation and Investment in the Canola Sector*, Wallingford: CABI Publishing.

Phillips, P., S. Smyth and W. Kerr (eds) (2006), *Governing risk in the 21st Century: Lessons from the World of Biotechnology*, Hauppauge, NY: Nova Science Publishers.

Picciotto, S. (2002), 'Introduction: reconceptualizing regulation in the era of globalization', *Journal of Law and Society*, **29**(1), 1–11.

Picciotto, R. (1995), 'Putting institutional economics to work: from participation to governance', World Bank discussion paper 304, accessed 15 August, 2005 at www-wds.worldbank.org/servlet/WDSContentServer/WDSP/IB/ 1995/09/01/ 000009265_3961219115337/Rendered/PDF/multi_page.pdf.

Polanyi, K. (1944/2001), *The Great Transformation: The Political and Economic Origins of Our Time*, Boston, MA: Beacon Press.

Polanyi, M. (1962), 'The republic of science: its political and economic theory', *Minerva*, **1**, 54–74.

Porter, M. (1980) *Competitive Strategy: Techniques for Analyzing Industries and Competitors*, New York: Free Press.

Porter, M. (1985), *Competitive Advantage: Creating and Sustaining Superior Performance*, New York: Free Press.

Porter, M. (1990), *The Comparative Advantage of Nations*, New York: Free Press.

Porter, M. (1998), 'Clusters and competition', in M. Porter (ed.), *On Competition*, Cambridge, MA: Harvard Business School Press, pp. 197–287.

Porter, M. (2000), 'Location, competition, and economic development: local clusters in a global economy', *Economic Development Quarterly*, **14**(1), 15–34.

Powell, D. (1999), 'Update: potential impacts of pollen from Bt-corn, November 15', accessed 17 November, 2006 at www.foodsafety network.ca/articles/705/bt-corn_impacts.pdf.

Powell, W. (1991), 'Neither market nor hierarchy: network forms of organizations', in G. Thompson, J. Frances, R. Lovacic and J. Mitchell (eds), *Markets, Hierarchies and Networks: The Coordination of Social Life*, London: Sage, pp. 265–76.

Powell, W. and S. Grodal (2005), 'Networks of innovators', in J. Fagerberg, D. Mowery and R. Nelson (eds), *The Oxford Handbook of Innovation*, Oxford: Oxford University Press, pp. 56–85.

Power, M. (1994), *The Audit Explosion*, London: Demos.

Putnam, R. (2000), *Bowling Alone: the Collapse and Revival of American Community*, London: Simon & Schuster.

Putnam, R., R. Leonardi and R. Nanetti (1993), *Making Democracy Work: Civic Traditions in Modern Italy*, Princeton, NJ: Princeton University Press.

Quist, D. and I. Chapela (2001), 'Transgenic DNA introgressed into traditional maize landraces in Oaxaca, Mexico', *Nature*, **414**, 541–3.

Rausser, G. and R. Goodhue (2003), 'Value differentiation', *Journal of Agricultural and Resource Economics*, **28**(3), 375–95.

Roelandt, T. and P. den Hertog (eds) (1999), *Boosting Innovation. The Cluster Approach. OECD Proceedings*, Paris: OECD.

Rhodes, R. (1995), *The New Governance: Governing Without Government*, The State of Britain Seminar II of a Joint Economic and Social Research Council/Regional Science Association seminar series, 24 January.

Ridley, M. (1999), *Genome: The Autobiography of a Species in 23 Chapters*, London: Fourth Estate.

Robbins, L. (1932), *An Essay on the Nature and Significance of Economic Science*, reprinted 1984, London: Palgrave Macmillan.

Rogers, E. (1983), *Diffusion of Innovations*, 3rd edn, New York: Free Press.

Rogers, E. (2003), *Diffusion of Innovations*, 5th edn, New York: Free Press.

Romer, P. (1990), 'Endogenous technological change', *Journal of Political Economy*, **98**(5:2), S71–S102.

Rosenau, J. (1995), 'Governance in the twenty-first century', *Global Governance*, **1**, 13–43.

Rosenberg, N. (1982), *Inside the Black Box: Technology and Economics*, Cambridge: Cambridge University Press.

Rosenberg, N. (1994), *Exploring the Black Box: Technology, Economics and History*, Cambridge: Cambridge University Press.

Rosenberg, N. (1997), 'Babbage: pioneer economist', accessed from the Babbage papers online at www.ex.ac.uk/BABBAGE/rosenb.htm.

Rosenfeld, S. (2002), 'Creating smart systems: a guide to cluster strategies in less favored regions', report generated for the Directorate General for Regional Policy and Cohesion of the European Commission, April, accessed 2 December, 2005 at http://europa.eu.int/comm/regional_policy /innovation/pdf/guide_rosenfeld_final.pdf.

Rosecrance, R. (1996), 'The rise of the virtual state', *Foreign Affairs*, **75**(4), 45–61.

Rousseau, J.J. (1762), *The Social Contract: or Principles of Political Right*, reprinted 2004, Kila, MT: Kessinger Publishing.

The Royal Society (2002), 'Genetically modified plants for food use and human health – an update', policy document 4/02, February, accessed 17 November, 2006 at http://image.guardian.co.uk/sys-files/guardian/ documents/2002/02/04/document-165.pdf.

Rugman, A. and J. D'Cruz (2000), *Multinationals as Flagship Firms: Regional Business Networks*, Oxford: Oxford University Press.

Rugman, A. and A. Verbeke (2001), 'Subsidiary-specific advantages in multinational enterprises', *Strategic Management Journal*, **22**(3), 237–50.

Ruttan, V. (2003), *Social Science Knowledge and Economic Development: An Institutional Design Perspective*, Ann Arbor, MI: University of Michigan Press.

Ryan, C. and P. Phillips (2005), *Using Social Network Analysis to Evaluate and Predict Performance: A Case-based Analysis of Genomics-based R&D Networks*, October, Zurich: Centre for Comparative and International Studies Applications in Social Network Analysis.

Salamon, L. (2002a), 'The new governance and the tools of public action: an introduction', in L. Salamon and O. Elliot (eds), *The Tools of Government: A Guide to New Governance*, Oxford: Oxford University Press, pp. 1–43.

Salamon, L. (2002b), 'The tools approach and the new governance: conclusions and implications', in L. Salamon and O. Elliot (eds), *The Tools of Government: A Guide to New Governance*, Oxford: Oxford University Press, pp. 1–43.

Salter, B. (ND), 'Transnational governance and cultural politics: the case of human embryonic stem cells and the European Union's Sixth Framework Program', working paper 1 of the project 'The global politics of human embryonic stem cell science', funded under the Economic and Social Research Council Stem Cell Initiative.

Salter, B. (ND), 'The global politics of human embryonic stem cell science', working paper 2 of the project 'The global politics of human embryonic stem cell science', funded under the ESRC Stem Cell Initiative.

Saviotti, P. (1996), *Technological Innovation, Variety, and the Economy*, Cheltenham UK and Brookfield, USA: Edward Elgar.

Saxenian, A. (1994), *Regional Advantage: Culture and Competition in Silicon Valley and Route 128*, Cambridge, MA: Harvard University Press.

Schumpeter, J. (1934), *The Theory of Economic Development*, Cambridge, MA: Harvard University Press.

Schumpeter, J, (1939), *Business Cycles: a Theoretical, Historical and Statistical Analysis of the Capitalist Process*, New York: McGraw-Hill.

Schumpeter, J. (1954), *Capitalism, Socialism, and Democracy*, London: George Allen and Unwin.

Scotchmer, S. (2004), *Innovation and Incentives*, Cambridge, MA: MIT Press.

Scott, C. (2002), 'Private regulation of the public sector: a neglected facet of contemporary governance', *Journal of Law and Society*, **29**(1), 56–76.

Seabright, P. (2004), *The Company of Strangers: A Natural History of Economic Life*, Princeton, NJ: Princton University Press.

Segarra, A. and J. Rawson (2001), 'Starlink™ corn controversy: background', CRS report for Congress RS20732, updated 10 January.

Shafaeddin, S. (2004), 'Who is the master? Who is the servant? Market or government?', United Nations Conference on Trade and Development discussion paper 2175, August, UNCTAD/OSG/DP/2004/9.

Shapin, S. and S. Shaffer (1985), *Leviathan and the Air-pump: Hobbes, Boyle, and the Experimental Life*, Princeton, NJ: Princeton University Press.

Shapiro, C. and H. Varian (1999), *Information Rules: A Strategic Guide to the Network Economy*, Boston, MA: Harvard University Press.

Sheldon, I. and T. Josling (2002), 'Biotechnology regulations and the WTO', International Agricultural Trade Research Consortium (IATRC) working paper No. 02-2.

Shrader-Frechette, K. (ND), 'Scientific method, anti-foundationalism and public decision making', in *RISK: Health, Safety & Environment*, accessed 15 September, 2005 at www.piercelaw.edu/risk/vol1/winter/Shrader.htm.

Shrader-Frechette, K. (1985), *Risk Analysis and the Scientific Method: Methodological and Ethical Problems with Evaluating Societal Hazards*, Boston, MA: D. Reidel Publishing Co.

Simon, H. (2000), 'Public administration in today's world of organizations and markets', *PS: Political Science and Politics*, **33**(4), 749–56.

Smith, M., *et al.* (2003), 'Historical perspectives on invention and creativity', *The Lemelson-MIT Program*, accessed 24 January, 2006 at www.web.mit.edu/invent/report.htm.

Smyth, S., P. Phillips, W. Kerr and G. Khachatourians (2004), *Regulating the Liabilities of Agricultural Biotechnology*, Wallingford: CABI Publishing.

Sorokin, P. (1959), *Social and Cultural Mobility*, New York: The Free Press.

Sorokin, P. (1959), *Basic Trends*, World Union Goodwill, December.

Spencer, B. and J. Brander. (1983), 'International R&D rivalry and industrial strategy', *Review of Economic Studies*, **50**, 707–22.

Steil, B., G. Victor and R. Nelson (eds) (2002), *Technological Innovation and Economic Performance*, Princeton, NJ: Princeton University Press.

Stein, J. (2001), *The Cult of Efficiency*, Toronto: Anansi and CBC.

Stern, S., M. Porter and J. Furman (2002), 'The determinants of national innovative capacity', *Research Policy*, **31**, 899–933.

Stewart, L. (1992), *The Rise of Public Science: Rhetoric, Technology, and Natural Philosophy in Newtonian Britain, 1660–1750*, Cambridge: Cambridge University Press.

Stigliz, J. (2002), *Globalization and its Discontents*, New York: W.W. Norton.

Stoker, G. (1998), 'Governance as theory: five propositions', *International Social Science Journal*, **50**(155), 17–28.

Storper, M. (1996), 'Institutions of the knowledge-based economy', in OECD (ed.), *OECD, Employment and Growth in the Knowledge-based Economy*, Paris: OECD, pp. 255–83.

Storper, M. (1997), *The Regional World: Territorial Development in a Global Economy*, New York: Guilford Press.

Strange, S. (1988), *States and Markets*, London: Pinter.

Strong, E. (1925), *The Psychology of Selling*, New York: McGraw-Hill.

Swartz, J. (2002), 'Internet firms fish for revenue from fees', *USA Today*, accessed 17 November, 2006, at www.usatoday.com/money/covers/2002-06-11-tech-sell.htm.

Tacket, C., H. Mason, G. Losonsky, J. Clements, M. Levine and C. Arntzen (1998), 'Immunogenicity in humans of a recombinant bacterial antigen delivered in a transgenic potato', *Nature Medicine*, **4**(5), 607–9.

Tacket, C., H. Mason, G. Losonsky, M. Estes, M. Levine and C. Arntzen (2000), 'Human immune responses to a novel Norwalk virus vaccine delivered in transgenic potatoes', *Journal of Infectious Diseases*, **182**, 302–5.

Tarschys, D. (2001), 'Wealth, values, institutions: trends in government and governance', in OECD (ed.), *Governance in the 21st Century*, Paris: pp. 27–44, accessed 17 November, 2006 at www.oecd.org/dataoecd/15/0/173944844.pdf.

Taylor, F. (1911), *The Principles of Scientific Management*, reprinted 1998, Mineola, NY: Dover Publications.

Tetley, W. (2000), 'Mixed jurisdictions: common law vs civil law (codified and uncodified)', *Louisiana Law Review* **60**, 677–738, accessed 25 October, 2005 at www.cisg.law.pace.edu/cisg/biblio/Tetley.html.

Thanavala, Y., M. Mahoney, S. Pal, A. Scott, L. Richter, N. Natarajan, P. Goodwin, C.J. Arntzen and H. S. Mason (2005), 'Immunogenicity in humans of an edible vaccine for hepatitis B', *Proceedings of the National Academy of Science*, **102**(9), 3378–82.

Thompson, Paul (ND), 'Risk objectivism and risk subjectivism: when are risks real?', in *RISK: Health, Safety & Environment*, accessed at www.piercelaw.edu/Risk/Vol1/winter/thompson.htm

Thornton, S. (2005), 'Karl Popper', *Stanford Encyclopedia of Philosophy*, accessed 26 August, 2005 at http://plato.stanford.edu/entries/popper/.

Tirole, J. (1988), *The Theory of Industrial Organization*, Cambridge, MA: MIT Press.

Truman, D. (1951), *The Governmental Process: Political Interests and Public Opinion*, London: Greenwood Press.

Valaskakis, K. (2001), 'Long-term trends in global governance: from "Westphalia" to "Seattle"', in OECD (ed.), *Governance in the 21st Century*, Paris: OECD, accessed 20 August, 2005 at www.oecd.org/dataoecd/15/0/17394484.pdf

Vogel, D. and R. Kagan (2004), *Dynamics of Regulatory Change: How Globalization Affects National Regulatory Policies*, Berkeley, CA: University of California Press.

Wahba, M. and L. Bridwell (1976), 'Maslow reconsidered: a review of research on the need hierarchy theory', *Organization Behavior and Human Performance*, **15**, 212–40.

Walras, L. (1874), *Elements of Pure Economics*, reprinted 2003, London: Routledge.

Warda, J. (1997), 'R&D tax incentives in OECD countries: how Canada compares', Conference Board of Canada members' briefing 190–97, January.

Wasserman, S. and K. Faust (1994), *Social Network Analysis*, New York: Cambridge University Press.

Weiss, T. (2000), 'Governance, good governance and global governance: conceptual and actual challenges', *Third World Quarterly*, **21**(5), 795–814.

Williamson, O. (1979), 'Transaction-cost economics: the governance of contractual relations', *Journal of Law and Economics* **22**, 233–61.

Williamson, O. (1981), 'The economics of organization: transaction cost approach', *American Journal of Sociology*, **87**, 548–77.

Williamson, O. (1983), 'Organization form, residual claimants, and corporate control', *Journal of Law and Economics*, **26**, 675–80.

Wolfe, D. and M. Gertler (2004), 'Clusters from the inside and out: local dynamics and global linkages', *Urban Studies*, **41** (5/6), 1071–93.

Woodley, B. (2001), *The Impact of Transformative Technologies on Governance: Some Lessons from History*, Institute on Governance, 25 October, accessed 11 August, 2005 at www.iog.ca/publications/transformative_tech.pdf.

Woodman, R.W., J.E. Sawyer and R.W. Griffin (1993), 'Toward a theory of organizational creativity', *Academy of Management Review*, **18**(2), 293–321.

World Bank, (1997), *World Development Report 1997: The State in a Changing World*, Washington, DC: The World Bank Group, accessed 11

May, 2006 at www.worldbank.org/html/extpb/wdr97/English/wdr97.com .htm

World Trade Organization (1994), *The Results of the Uruguay Round of Multilateral Trade Negotiations: The Legal Text*, Geneva: GATT.

WTO Appellate Decision (1998a), *Australia – Measures Affecting Importation of Salmon – Complaint by Canada*, AB-1998-5, appeal from WT/DS18, accessed at http://wto.org

WTO Appellate Decision (1998b), *EC Measures Concerning Meat and Meat Products (Hormones) – Complaint by the United States and Canada*, AB-1997-4, appeal from WT/DS26/R/USA and WT/DS48/R/CAN, accessed at http://wto.org.

WTO Appellate Decision (1999), *Japan – Measures Affecting Agricultural Product – Complaint by United States*, AB-1998-8, appeal from WT/DS76, accessed at http://wto.org

Wright, R. (2004), *A Short History of Progress*, Toronto: Anansi.

Zucker, L., M. Darby and M. Brewer (1998), 'Intellectual human capital and the birth of US biotechnology enterprises', *American Journal of Economics*, **88**(1), 290–306.

# Index